MULTIPLE STRESSORS: LITERATURE REVIEW AND GAP ANALYSIS

by:

Stella M. Swanson, Ph.D.
Christian G. J. Fraikin, D.E.C.
David J. Van Den Assem
Golder Associates Ltd.

Joseph M. Culp, Ph.D.
Richard B. Lowell, Ph.D.
National Water Research Institute

Susan Eipert, Ph.D.
Eipert Information Services

Gary A. Lamberti, Ph.D.
University of Notre Dame

Barry R. Taylor, Ph.D.
Taylor Mazier Associates

2004

WERF

The Water Environment Research Foundation, a not-for-profit organization, funds and manages water quality research for its subscribers through a diverse public-private partnership between municipal utilities, corporations, academia, industry, and the federal government. WERF subscribers include municipal and regional water and wastewater utilities, industrial corporations, environmental engineering firms, and others that share a commitment to cost-effective water quality solutions. WERF is dedicated to advancing science and technology addressing water quality issues as they impact water resources, the atmosphere, the lands, and quality of life.

For more information, contact:
Water Environment Research Foundation
635 Slaters Lane, Suite 300
Alexandria, VA 22314-1177
Tel: (703) 684-2470
Fax: (703) 299-0742
www.werf.org
werf@werf.org

This report was co-published by the following organization. For nonsubscriber sales information, contact:

IWA Publishing
Alliance House, 12 Caxton Street
London SW1H 0QS, United Kingdom
Tel: +44 (0) 20 7654 5500
Fax: +44 (0) 20 7654 5555
www.iwapublishing.com
publications@iwap.co.uk

Library of Congress Catalog Card Number: 2004103670

IWAP ISBN: 1-84339-719-6.

Golder Associates Ltd., Taylor Mazier Associates, National Water Research Institute, Eipert Information Services, University of Notre Dame

The research on which this report is based was funded, in part, by the United States Environmental Protection Agency through Cooperative Agreement No. CR-827345-01-1 with the Water Environment Research Foundation (WERF). Unless an EPA logo appears on the cover, this report is a publication of WERF, not EPA. Funds awarded under the Cooperative Agreement cited above were not used for editorial services, reproduction, printing, or distribution

ACKNOWLEDGMENTS

Supplemental support from the U.S. Environmental Protection Agency is gratefully acknowledged. Members of the Project Subcommittee were involved in two planning sessions, providing lively discussion and valuable contributions to the direction and content of the report both in these sessions and throughout the duration of the project. Their contributions are greatly appreciated. The project team also thanks Mr. David O'Sadnick and Ms. Shannon Poppe of Golder Associates in Denver for hosting the year one planning session.

Report Preparation

Principal Investigator:
Stella M. Swanson, Ph.D.
Golder Associates Ltd.

Project Team:
Joseph M. Culp, Ph.D.
National Water Research Institute

Susan Eipert, Ph.D.
Eipert Information Services

Christian G.J. Fraikin, D.E.C.
Golder Associates Ltd.

Gary A. Lamberti, Ph.D.
University of Notre Dame

Richard B. Lowell, Ph.D.
National Water Research Institute

Barry R. Taylor, Ph.D.
Taylor Mazier Associates

David J. Van Den Assem
Golder Associates Ltd.

Project Subcommittee

Scott Dyer, Ph.D., Chair
Procter & Gamble

Michael V. Bastian
CH2M HILL

Timothy J. Hall
National Council for Air and Stream Improvement, Inc.

Carolyn T. Hunsaker, D.Env.
U.S.D.A. Forest Service

Dixon H. Landers, Ph.D.
U.S. Environmental Protection Agency

Irwin Polls
Metropolitan Water Reclamation District of Greater Chicago

Kent W. Thornton, Ph.D.
FTN Associates

Water Environment Research Foundation Staff

Director of Research: Daniel M. Woltering, Ph.D.
Project Managers: Dean Carpenter, D.Env.
Eugenia Bruno Hart

ABSTRACT AND BENEFITS

There is a pressing need for developing and testing a general set of theories to provide a confident basis for prediction of multiple stressor effects. Confident prediction is central to confident decision making in water pollution control. Consequently, this research was to provide a study design based on good science that helps establish a general, conceptual approach to multiple stressors. The first step of the process was to conduct a review of the literature on experimental studies of multiple-stressor effects in aquatic ecosystems.

This literature review provides a critique of the existing body of knowledge concerning multiple stressors, identifies knowledge gaps, and provides a searchable, annotated bibliography of multiple-stressor research with nearly 250 entries (available upon request). Stressors have been sorted into four main groups for the purpose of presentation: metals, organics, water quality (nutrients, chlorine, temperature, and similar physical or chemical variables), and disturbance. The review of each of these four groups provides a synopsis and review in the context of risk assessment. The synopsis includes the stressors, receptors, and measures included in each of the reviewed studies. The contribution of the current literature to the construction of a generalized conceptual model of multiple-stressor effects is also reviewed.

Benefits:

- Provides a searchable annotated database of literature (available upon request) pertaining to experimental studies of multiple stressors and a readily accessible source of experimental data for use in understanding how multiple stressors interact in ecosystems.
- Places multiple-stressor research into a risk assessment context.
- Includes a discussion of the interpretation of impaired streams, processes governing ecosystem response to stress, and the nature of recovery, and return to some desired state of the ecosystem.
- Compares and contrasts observational and experimental studies of multiple stressors.
- Evaluates the contribution of the existing body of multiple-stressor research to the development of a conceptual model for multiple-stressor interactions and effects in ecosystems.
- Identifies the primary gaps in the multiple-stressor literature,
- Provides the basis for the conceptual model and study design presented in the companion report, "Multiple Stressors: Risk-Based Framework and Experimental Design for Cause/Effect Relationships" (WERF Report 00-ECO-2A).

Keywords: stressor interaction, ecosystem response, pattern, process, risk assessment, mesocosm, critique, data gaps, conceptual model

TABLE OF CONTENTS

LIST OF TABLES

LIST OF FIGURES

LIST OF ACRONYMS

2,4-D	2,4-Dichlorophenoxyacetic Acid
2,4,5-T	2,4,5-Trichlorophenoxyacetic Acid
ATP	Adenosine Triphosphate
BEAST	Benthic Assessment of Sediment
BKME	Bleached Kraft Mill Effluent
BOD	Biochemical Oxygen Demand
DOC	Dissolved Organic Carbon
DO	Dissolved Oxygen
ERA	Ecological Risk Assessment
IBI	Index of Biotic Integrity
MERL	Marine Ecosystems Research Laboratory
PA	Phosphoarginine
PAH	Polycyclic Aromatic Hydrocarbons
PCB	Polychlorinated Biphenyls
QSAR	Quantitative Structure-Activity Relationships
RivPacs	River Invertebrate Prediction and Classification Scheme
SETAC	Society of Environmental Toxicology and Chemistry
TMDLs	Total Maximum Daily Loads
U.S. EPA	United States Environmental Protection Agency
U.S. FWS	United States Fish and Wildlife Service
WERF	Water Environment Research Foundation

EXECUTIVE SUMMARY

ES.1 Introduction

There is a pressing need to better understand the effects of multiple stressors in stream ecosystems. Several organizations, including the Water Environment Research Foundation (WERF), have contributed to the process of understanding the effects of multiple stressors by publishing a number of scientific reviews, suggested frameworks, and compendia of available tools, including models, statistical approaches, and effect-based assessment methods. However, further progress toward understanding and predicting multiple-stressor effects depends on the development and testing of a general set of multiple-stressor theories. Confident prediction leads to confident decisions for water pollution control.

The merits of developing a conceptual model for multiple stressors that can be experimentally tested and confirmed in the field were identified by WERF in 1999. WERF subsequently commissioned this study to provide a science-based study design that helps establish a general, conceptual approach to multiple stressors. The first step in meeting the goal was to review the literature pertaining to experimental studies of multiple stressors. The objectives of the research are as follows:

- Review and critique the existing body of knowledge for multiple stressors
- Develop a searchable, annotated bibliography of multiple-stressor research
- Identify gaps in the body of knowledge
- Develop a flexible and scientifically defensible conceptual model of the environmental effects of multiple stressors on river and stream ecosystems
- Develop a study design to test the conceptual model of multiple-stressor effects
- Document the process used to arrive at the conceptual model

This report addresses the first three objectives. It includes the review and critique of the existing body of multiple-stressor research and identifies the gaps in this research. The annotated bibliography has been provided to WERF as a searchable database (available upon request). Developing the conceptual model, study design and documenting the process used to arrive at the conceptual model are addressed in a companion report, "Multiple Stressors: Risk-Based Framework and Experimental Design for Cause/Effect Relationships" (00-ECO-2A).

The study team conducted a simultaneous search of over 30 databases, including all of the major sources for published work and government literature. The search produced a list of approximately 600 titles. The study team screened the titles and deleted irrelevant titles.

After screening, the 386 remaining references were distributed among the team members for review. This served a twofold purpose: to determine which papers were sufficiently relevant and useful to be included in the database, and to provide annotation, including a quality ranking for the papers in question. Approximately 50 of the candidate titles from the literature search turned out to be irretrievable. These were all conference abstracts or unpublished reports. Of the 336 reviewed papers, 178, or approximately 53%, were rejected as unsuitable by the reviewers. New papers have since been added. Consequently the database currently contains 242 relevant, good quality articles. The small number of acceptable papers clearly indicates that the effect of multiple stressors is a relatively new area of research. All of the relevant articles have been compiled into an annotated database that is fully searchable by any of the keyword fields and is available from WERF by request.

ES.2 Patterns and Process of Ecosystem Response to Multiple Stressors

The assessment of multiple-stressor effects can be based on two fundamental approaches. These two approaches are the search for:

- Pattern—observational evidence of a response pattern that is correlated with environmental stressors
- Process—experimental evidence of a cause/effect relationship between responses and particular, measured combinations of stressors

This literature review largely considers "process" studies because previous projects have provided comprehensive reviews of observational studies and the search for pattern (Rosenberg and Resh 1993; Fairbrother and Bennett, 2000; Harwell and Gentile, 2000). However, the overall framework of both pattern and process was useful when developing the Study Design for multiple stressors. Therefore, a brief overview of the search for pattern using observational studies is included in the review.

ES.2.1 The Search for Pattern

The search for pattern is usually field-based and observational. All of the "search for pattern" approaches are based on the premise of "letting the system tell you about its current state and deducing how it got there" (Harwell and Gentile, 2000). This approach can include the statistical analyses of relationships between biological response variables (such as benthic invertebrate community metrics) and measured stressors. Various statistical tools are available for these analyses, but the objective is always to partition the variance of observed responses among the various stressors to see which stressors, either alone or in combination, can explain the observed responses (Fairbrother and Bennett, 2000). Integrated response variables (sometimes called biocriteria, biotic indices, or metrics) are a commonly used tool in observational studies. The relationships between these metrics and stressors have been the subject of numerous publications because of the popularity of their use for the identification of impaired aquatic systems.

The review of the "search for pattern" literature showed that observational evidence is an important part of developing our understanding of multiple-stressor effects; however, it cannot be the sole basis for the assessment of risk. Teasing apart the effects of multiple stressors will require a theoretical foundation that will foster predictions about how stressors will interact. This is because the search for pattern through the use of community metrics, the reference condition, multimetrics such as rapid bioassessment (Barbour, 1997) or effects-based monitoring (Environment Canada, 1998) provides little diagnostic or predictive power. Formal stressor identification methods, although rigorous and based on established and accepted weight-of-evidence approaches, cannot provide a theoretical foundation for the prediction of ecosystem responses to multiple stressors.

ES.2.2 The Search for Process

The search for the processes that govern responses to multiple stressors requires an experimental approach. The experimental approach requires hypotheses to test, which are often generated by observational data (i.e., the experiments are part of follow-up studies used to confirm cause/effect relationships). However, the generation of hypotheses should go beyond a case-by-case confirmation of putative cause/effect relationships and extend to generalizations that are useful for prediction.

The literature review team searched for studies with present multiple-stressor hypotheses. However, multiple-stressor literature with hypotheses was scarce, so the study team decided to include experimental studies that involving more than one stressor, even if no hypotheses were presented. It was hoped that the experimental studies of multiple stressors would reveal some consistent responses, at least within certain stressor classes. It was also hoped that the experimental studies would indicate whether the commonly used assumption of additivity of multiple-stressor effects was often, seldom, or never valid.

ES.3 Risk Assessment as the Framework for the Literature Review

The ecological risk assessment (ERA) method currently in use in the United States and Canada is a useful framework for the review of multiple-stressor research. One primary advantage of the risk assessment framework is the structure it imposes on the analysis and interpretation of exposure and effects data. The risk assessment framework requires a clear, concise definition of the problem before proceeding to data collection, modeling, and analysis. The problem definition stage, when done rigorously, provides a clear conceptual model that generates hypotheses regarding the relationship between stressors and ecological endpoints. The problem definition stage also requires an explicit expression of the ecological assessment endpoints with an accompanying list of measures that can be used to assess whether those assessment endpoints will be met. Thus, decision makers will have results that clearly relate to ecological management goals that integrate ecological and societal values.

The multiple-stressor literature was reviewed for its contribution to conceptual models that present hypothesized relationships between multiple stressors and ecological endpoints. The literature was also reviewed for its use of assessment endpoints, the selected measures of effects, and the relevance of these endpoints and measures to populations, communities, and ecosystems.

ES.4 Theories of Ecosystem Response

Professional judgement plays an important role in risk characterization, therefore, it is imperative that the theory or model used to interpret risk is, at the least, referenced and preferably, explained. Several useful models of the responses of ecosystems to stress exist, along with a number of methods for developing a credible risk characterization.

Odum et al.'s (1979) ecosystem subsidy-stress model is one of the best known models for predicting the response of ecosystems to stressors. Their subsidy-stress perspective holds that the low levels of some stressor can "subsidize" an ecosystem process but that further loading beyond some threshold will "stress" that same ecosystem. Subsidies are assumed to have positive effects on the ecosystem, whereas stresses are negative. Although Odum's predictions are based on the simple assumption that undisturbed ecosystems develop along an expected successional pathway, they serve as a valuable tool for developing hypotheses regarding structural and functional responses in disturbed ecosystems.

The choice of structural or functional attributes of ecosystems as measures of stress has been a subject of debate. Odum (1985) and others have argued that structural elements of ecosystems (e.g., community structure, biomass) are more sensitive indicators of anthropogenic stress than are process-oriented functional attributes, because species replacement and other adjustments keep the overall function of the ecosystem steady until the stress is overwhelming (Schindler, 1987). However, Pratt and Cairns (1996) challenge this view and cite examples demonstrating the sensitivity of functional ecosystem parameters. Species-specific approaches also may be most suitable for particular circumstances, such as when threatened or endangered species are the focus. In summary, it is doubtful that structural or functional endpoints alone will be suitable for all cases, but rather a pluralistic approach will be needed for evaluating system responses to environmental stressors.

A key to Odum's approach is the assumption that a single major stressor has disrupted the ecosystem from its "nominal" state, thereby producing either a subsidy or stress response. The notion of "recovery" to some nominal state is a pervasive and important underlying concept that affects the development of conceptual models of multiple-stressor effects in ecosystems. The literature on ecological resilience is relevant to this discussion. A broad definition of resilience is "the capacity of a system to absorb disturbances" (Raufflet, 2000).

The definition of recovery is central to the understanding of the resilience of an ecosystem. The question is "recovery to what state?" The literature emphasis on resiliency of ecosystem function often assumes that recovery will be to a previous equilibrium state where productivity returns to a balance with available resources and mineral cycling is maintained. Some researchers with this point of view state that ecosystems are not as sensitive as the most sensitive species because of the redundancy in roles and functions provided by the diversity of species. Such an approach assumes that effects are reversible. The literature emphasis resiliency of ecosystem structure stresses the importance of the conservation of biodiversity. Researchers have noted that although ecosystems have adapted to natural perturbations such as fire, the response to cumulative effects of stressors such as climate change, increased ultraviolet exposure because of ozone depletion, and acidification can produce unexpected and sometimes severe effects. Changes in community structure (species richness, proportion of r-strategists, size of organisms, lifespan of organisms, and length of food chains) can affect habitat and food niche availability, as well as functional processes.

Recently, the whole concept of recovery and reversibility has been challenged, at least with respect to the idea that ecosystems can return to a prestressed equilibrium state. These authors (Matthews et al., 1996; Landis et al., 2000; Sandberg and Landis, 2001) contend that ecological communities tend to preserve information about every event in their history and they are not at equilibrium, but rather are in the process of responding to their own unique history of interactions. The theory of "community conditioning" suggests that ecological systems are complex, nonequilibrium systems and that our assumption that ecosystems can be "managed" to preserve or return to some prestressed state is spurious. Instead, ecosystem management must recognize that the postdevelopment state of the ecosystem will be different—it may be equally functional with the desired structure, but it will not be the same as before development.

Although the assertion that ecosystems do not return to some previous state is intriguing and has an experimental and theoretical basis, it tends to cause a certain amount of paralysis when one is faced with the requirement to make decisions about mitigation or remediation measures. How does one assess whether a certain amount of impairment requires remediation if we do not know where the ecosystem may be heading? How do we measure our success in remediation against a backdrop of constant change along a unique trajectory? Somehow, we still need some sort of measure of the "state of the ecosystem" versus some idea of the "desired state." The desired state may be as straightforward as a statement of the desired human uses for the stream and the conditions that would allow those uses.

The authors of this review believe that there is a middle ground between the rigid adherence to Clementian ideas of ecosystem recovery (Clements, 1916, 1936) and the appealing, but difficult-to-implement ideas about community conditioning. However, the definition of that middle ground will be difficult or impossible unless we make progress in our fundamental understanding of how particular combinations of stressors may be expected to behave and how particular ecosystems may be expected to respond. Hence, the central focus of the literature review was on studies that contributed to our understanding of the relationship between particular combinations of multiple stressors and ecosystem response.

ES.5 Review and Identification of Gaps in the Literature

ES.5.1 Gaps in the Theoretical Basis for Multiple-Stressor Effects

A rigorous, theoretical basis for the prediction of multiple-stressor effects does not exist. Despite the wealth of observational data, the existence of several useful tools for interpretation of cause/effect relationships [including formal stressor identification methods], and the studies reviewed in this report, there are no tools that allow a confident, *a priori,* prediction of ecosystem response to multiple stressors. Retrospective studies can use weight-of-evidence approaches and the standard rules for establishing cause/effect relationships to piece together a plausible story regarding the stressors most likely to have caused the observed responses. However, these retrospective studies are time consuming, data intensive, and tend to result in analyses that cannot be readily extrapolated to other cases. Furthermore, uncertainty associated with these retrospective studies still centers around model error (i.e., our imperfect understanding of how multiple stressors can be expected to interact and how ecosystems can be expected to respond). Instead, we default to a standard assumption of additivity (in the case of chemical stressors) or to a case-by-case, professional process.

The ultimate goal for the assessment of risks from multiple stressors is to have an understanding of multiple-stressor effects that is sufficiently robust to support decision making.

The primary outcomes would be credible, defensible, and practical conceptual models that can be used as the starting point for risk assessments at a wide variety of sites.

Much of the multiple-stressor literature presents an inventory of stressors with no study of the interactions between the stressors. The interactive effects of these stressors are not addressed, largely because it is difficult to apportion the effects observed to the variety of stressors present. Studies that use the weight-of-evidence case for cause/effect relationships among stressors and receptors make an attempt to address these interactive effects.

There is considerable difficulty in quantifying the effects of multiple stressors that interact in a nonlinear manner. With the benefit of extensive research on specific interactions, we can produce reasonable predictions of cumulative stress (e.g., from many years of study of aluminum plus pH). It is a greater challenge to calculate the cumulative stress from two stressors whose interaction has not been thoroughly studied.

The current literature provides, at best, a series of site-specific glimpses of the response of ecological communities and ecosystems to multiple stressors. The definition of the problem provided by authors is usually very focused on a particular site or a particular combination of stressors. There is seldom, if ever, any reference to a more generalized model of multiple-stressor effects apart from the discussion of the expectations regarding additivity versus synergism. Many articles that purport to be multiple-stressor studies do not go beyond an inventory of the various stresses and upsets affecting the ecosystem or attempt to assess the interactions among them.

The exceptions to these studies are papers that define the problem in terms of building a weight-of-evidence case for cause/effect relationships among stressors and receptors, or papers that refer to a suggested general response pattern to stressors (e.g., Munkittrick and Dixon, 1989; Gibbons and Munkittrick, 1994). However, even these papers seldom place the definition of the problem in a wider framework provided by a general model such as Odum's model of ecosystem stress (Odum, 1985).

One notable paper by Lewis (1999) presents three testable hypotheses concerning general responses of ecosystems to stress. They are: 1) a given biotic compartment in an incrementally stressed ecosystem will show suppression of biodiversity first and suppression of biomass subsequently (similar to the idea expressed by Schindler, 1987); 2) suppression of ecosystem function is primarily caused by reduced biomass, rather than loss of diversity itself; and 3) suppression of biomass differs among biotic compartments according to a predictable series of thresholds (i.e., multicellular organisms are more sensitive than unicellular autotrophic algae which are more sensitive than microbial decomposers). The third hypothesis implies a predictable sequence of ecosystem processes that will be lost as the severity of stress increases. Lewis's work is seminal in that it goes beyond simply listing the responses of a particular system to a particular set of stressors, but instead provides a general, testable model of ecosystem behavior. Unfortunately, work to test these hypotheses has only just begun, although Lewis (1999) suggests that preliminary studies on streams suffering acid mine drainage are consistent with his hypotheses.

Because of the lack of a broader context to multiple-stressor research, the current literature does not provide a readily accessible basis for the construction of generalized conceptual models of multiple-stressor effects on aquatic ecosystems. However, a few tentative generalizations may be possible based on research to date and extrapolation from broader ecological theory.

ES.5.2 Generalizations Derived from the Literature

Some tentative generalizations resulted from the literature review. Generally, it appears that stresses of most kinds have a simple additive effect on *individual organisms*, but the response to stress is nonlinear. The organizational level that is being studied will determine the type and additivity of effects. However, we can anticipate that populations and communities will respond to stress in a predictable, linear fashion (above a lower threshold) because increasingly less-sensitive individuals or species will be removed as the intensity of the stressor(s) increases. The effect of multiple stressors is probably more often additive than synergistic or antagonistic at the population level.

According to Odum (1985), stressed ecosystems maintain processes at the expense of species; consequently stressors may have sharp thresholds of effect, below which the effect is minor or even undetectable (still enough species left to function). Above this level the effect is abruptly pronounced (no longer enough species or individuals to support essential processes). If the stressor is stimulatory of growth or production, like heat or nutrients, then a subsidy effect may be observed at low levels (Odum, 1985). The subsidy side of the equation in Odum's model has been given less consideration than the stress side. Nevertheless, scattered observations in the literature broadly support Odum's conception. Even toxic chemicals may act as subsidies to the ecosystem if the concentration is low enough. The nature of interactions between stressors and subsidies will determine the overall ecosystem response to multiple stressors.

Lastly, it is important to distinguish the interactions between the stressors from the interactions between stressor effects. Much of the confusion about interactions between broadly different stressors arises from a failure to make this distinction. Complex and apparently daunting stressors (i.e., climate change, invasive species, overharvesting, and flow regulation) can perhaps be rendered tractable by carefully and painstakingly considering their effects on ecosystems. The risk assessment can concentrate on precisely explicated changes to the system. There is a lack of both theory and tools for dealing with the so-called "Class 4" type of multiple-stressor problems. A Class 4 problem is one with multiple classes of stressors (chemical, physical, and biological) from multiple sources with interactions acting on multiple ecological receptors (Harwell and Gentile, 2000).

ES.5.3 Data Gaps in the Context of Risk Assessment

The literature review found that the most common combinations are metals or metals plus other stressors. A large volume of literature was available on the interactions of metals and pH. Studies of pesticides with other stressors are more common than studies just involving multiple pesticides, but experimental studies of combinations of organic compounds other than pesticides are very rare. Most organic compound multiple-stressor studies involved PCBs, and very few involved PAHs. There is a notable lack of experimental studies of combinations of stressors associated with agriculture, except for uncharacterized mixtures such as "agricultural pollution."

Few studies have been found on the effects of the four stressors identified as "top priority" by the U.S. EPA Integrated Risk Project (i.e., climate change; flow regulation; invasion of exotic species; and habitat alteration).

Multiple-stressor combinations used in experimental studies are summarized in Table ES-1.

Table ES-1. Number of Experimental Multiple-stressor Studies by Stressor Category.

Stressor Category	Number of Studies
metals	30
metals plus other stressors	20
pesticides	7
pesticides plus other stressors	12
organic compounds plus other stressors	23
nutrients/major ions plus other stressors	26
habitat change/invasion of nonindigenous species	6

Although the importance of building a relationship between stressors and receptors is of great importance, so too is the correct choice of receptors to study. The usefulness (or lack thereof) of the selected receptors as suitable indicators of the community or ecosystem is seldom critically examined in the literature.

The majority of studies discussed single species receptors, or interactions between two species under the presence of stressors. Few studies examined multispecies receptors, unless they were community assemblages of aufwuchs (plants and animals adhering to parts of rooted aquatic plants). Whole community studies were very rare, likely caused by the level of complexity inherent in larger systems.

Metal studies focused mostly on fish species. Papers studying the effects of multiple stressors on periphytons were noticeably absent.

Studies of the interactions of metals and other stressors showed a more even division between fish, benthic invertebrates and algae, although aquatic macrophyte studies were under-represented in the literature.

Standard test species were rarely used in studies of organic compounds in combination with other stressors. There was seldom any explanation of why the particular receptors were chosen. In some cases, receptors were taken from the study areas to the laboratory or field mesocosm. The choice of receptors was therefore implied to be a representation of the receptors found in the study lake, river, or stream

The current literature relies primarily on the traditional measures of exposure in aquatic systems (i.e., concentrations in water and/or sediments). These measures are sometimes supplemented with measures of tissue residues. There is insufficient information relating tissue residues to effects, despite the fact that measurement of stressor concentration in tissue would provide an estimate of "dose" and would reduce the uncertainty associated with estimates of bioavailability.

The issues regarding exposure assessment for multiple stressors also include:

- The rationale for the selection of exposure pathways in multiple-stressor research is rarely put in a broader context that would contribute to a generalized model of "critical pathways" that may exist when more than one stressor is present;
- There is a need for a wide-ranging exercise that produces hypothesized critical pathways in the presence of the top-priority stressors. This exercise may reveal unique critical pathways that would not have been thought about before; and
- When multiple-stressor studies combine chemical and nonchemical stressors, the "cumulative exposure" to these stressors is difficult to describe. The current state of the science is a simple listing of the chemical exposure (as concentration or tissue residue) and the relative changes in nonchemical stressors (such as relative change in habitat).

No unique measures of effects that would contribute to our understanding of the mechanisms underlying the toxicology of metal mixtures were observed in the literature. Measures of effects on individuals in experimental studies are either standardized single species toxicity test measures or measures specific to the toxic mechanism. Measures of effects on populations and communities in experimental studies are either standardized single species toxicity test measures or measures specific to the toxic mechanisms.

Effects measures relevant to populations (e.g., birth rate, recruitment, mortality rate) require development. The sensitivity of community-level measures and the usefulness of these measures in testing hypotheses regarding multiple-stressor interactions require more critical evaluation.

Measures useful in distinguishing additive, antagonistic or synergistic effects require identification and development. Traditional single-organism or community measures may not be the most informative regarding multiple-stressor effects because of high variability, or the lack of any connection with the mechanisms of multiple-stressor interactions.

ES.5.4 Assessment Endpoints and Conceptual Models

There are no examples of assessment endpoints for experimental studies of multiple stressors in the literature. Conceptual models are also missing, except for the occasional mention of the expectation of additive versus synergistic effects.

Assessment endpoints must be measurable ecosystem characteristics that adequately represent management goals (U.S. EPA, 1998). For example, if the management goal is to reduce the risk to striped bass populations from multiple-stressor interactions to acceptable levels, then the assessment endpoint must be sensitive to the identified stressors, yet respond in different ways to the different stressors. Striped bass populations may be adversely affected at several life stages, in different habitats, through different ways and by different stressors. A useful assessment endpoint would stipulate the attributes of striped bass populations most likely to respond to the particular combination of stressors under investigation. This assessment endpoint would be accompanied by a set of measures that encompass the range of responses of the various life stages to the different stressors.

Conceptual models consist of two main parts, a set of risk hypotheses that describe predicted relationships among stressor, exposure, and assessment endpoint response, and a diagram that illustrates the relationships in the risk hypotheses (U.S. EPA, 1998). Risk hypotheses are specific assumptions about potential risk to assessment endpoints and may be based on theory and logic, empirical data, mathematical models, or probabilistic models (U.S. EPA, 1998). There was a notable lack of risk hypotheses meeting these requirements in the literature. Conceptual model diagrams are a visual representation of these risk hypotheses. There was also a notable lack of conceptual model diagrams in the literature.

The lack of a multiple-stressor conceptual model is a primary source of uncertainty in the assessment of multiple-stressor risks. Because of this lack, the risk characterization of multiple stressors may misinterpret the actual risks.

"Multiple stressors are almost always present, and can be the source of confusion, especially when conceptual models focus strictly on single stressors" (U.S. EPA, 1998). Except for additivity, this notion is seldom addressed explicitly in the literature. Any contribution to the development of general conceptual models for multiple-stressor situations will be very valuable and was a particular focus of the accompanying report to this literature review.

ES.6 Conclusions

The review of the literature supports the need for an experimental approach to multiple-stressor effects. The generation of testable hypotheses is probably the most useful aspect of observational research in multiple stressors. However, observational studies are not sufficient to provide a sound theoretical basis for understanding multiple-stressor effects in aquatic ecosystems. Although observational studies have been useful in identifying the general attributes of systems exposed to multiple stressors, the actual role of particular stressors versus the role of the mixture of stressors is difficult to deduce from these data. Furthermore, the ecological processes leading to the observed responses can only be speculated on. Thus, the ability to predict the response of, for example, a periphyton or benthic invertebrate community is limited.

An experimental approach is required to test hypotheses arising from conceptual models of multiple-stressor effects. Experimental results can be used in combination with observational field studies in an iterative manner to produce increasingly robust conceptual models.

The results of the literature review were used to design a conceptual model of multiple-stressor effects that will lead to the development of an experimental approach. The accompanying study design document (WERF Report 00-ECO-2A) includes provision for complementary field studies; however, it emphasizes the experimental approach because of the demonstrated lack of a sound experimental basis for multiple-stressor theories.

CHAPTER 1.0

INTRODUCTION

1.1 Objectives

This project develops a report and annotated bibliography (available upon request) that demonstrates an in-depth understanding of the literature on multiple stressors as studied in experimental stream systems (i.e., mesocosms and field experiments—including human-derived and natural events) as they pertain to multiple stressors. The report also identifies and rates important gaps in the literature regarding multiple-stressor research.

Using a risk assessment framework, the companion report (WERF 00-ECO-2A) designs a multiyear field- and/or mesocosm-level research study that will identify a potential site for the study; categorize the types of ecosystem responses to stressors; evaluate relationships among multiple stressors and ecosystem responses; synthesize information concerning the cumulative effects of stressors; and measure recovery processes following the removal of one or more stressors.

1.2 Definitions

Two major sources of terminology are used in relation to multiple stressors—risk assessment and impact assessment. These two closely related fields have succeeded in generating separate sets of terms that are often nearly, but not totally, synonymous. To avoid creating further confusion and controversy, we provide definitions for frequently used terms in the glossary. The definitions are largely derived from the risk assessment lexicon because the literature review was performed within a risk assessment framework.

1.3 Focus of the Literature Review

The focus of this report is the true multiple-stressor literature. To be included in the literature review, a multiple-stressor study must consider more than one stressor operating at the same time and within the same space. It can include, but should not be limited to, the relationships between a particular stressor and some environmental modifiers (e.g., a metal and water hardness). It must explicitly measure and account for responses to identified and measured stressors. It cannot be a study of responses to a generalized suite of watershed characteristics that are described but not quantified.

This restrictive definition of the multiple-stressor literature was based on the underlying requirement that the study must contribute to our ability to assess and predict the risk from particular multiple-stressor combinations.

1.4 The Organization of the Report

The second chapter of this report discusses the conceptual framework for the literature review. It separates the multiple-stressor literature into two general categories ("pattern" and "process"). It then reviews risk assessment procedures, because the literature is reviewed for its contribution to assessing the risks of multiple stressors. Finally, it presents an overall theory of the response of ecosystems to stress and explains the use of the weight-of-evidence approach. The ecosystem subsidy-stress theory can assist the characterization of risk from multiple stressors. The weight-of-evidence approach is required if a rigorous and comprehensive understanding of multiple stressors is to be obtained.

Chapter 3.0 briefly describes the methods used to locate and compile the multiple-stressor literature. Chapter 4.0 presents the review of the multiple-stressor literature within the risk assessment context. The literature review combines stressors into four main groups for ease of presentation. These groups are:

- Metals: mixtures of metals plus metals combined with other stressors.
- Organics: mixtures of, or including, organic compounds.
- Water Quality: nutrients, chlorine, temperature, and similar physical or chemical variables.
- Disturbance: habitat change, species invasions, flow regulation, harvesting, and climate change.

Chapter 5.0 presents an evaluation of the multiple-stressor literature in terms of how it contributes to the development of a conceptual model of multiple-stressor effects. This chapter also provides some alternative frameworks for the assessment of multiple stressors and ends with an illustration of the application of the weight-of-evidence approach to the study of multiple stressors. Chapter 6.0 provides an overview of the theory of multiple-stressor effects and then reviews mesocosm methods.

The final chapter, Chapter 7.0, identifies the primary gaps in our understanding of multiple stressors. Chapter 7.0 also provides a brief overview of how the findings of the literature review link to the study design report.

CHAPTER 2.0

CONCEPTUAL APPROACH TO THE LITERATURE REVIEW

2.1 Pattern and Process of Response to Multiple Stressors

The assessment of multiple-stressor effects can be based on two fundamental approaches. These two approaches are the search for:

- Pattern (i.e., observational evidence of a response pattern that is correlated with environmental stressors); and
- Process (i.e., experimental evidence of a cause/effect relationship between responses and particular, measured combinations of stressors).

This literature review largely considers "process" studies because previous WERF projects have provided comprehensive reviews of observational studies and the search for pattern (Gerritsen and Jessup, 2004; Paulson et al., 2001). However, the overall framework of both pattern and process will be useful when developing the study design for multiple stressors (Year Two of this project) because of the utility of the weight-of-evidence approach. Therefore, a brief overview of the search for pattern using observational studies is provided below.

2.2 The Search for Pattern

The search for pattern is usually field-based and observational. All of the "search for pattern" approaches are based on the premise of "letting the system tell you about its current state and deducing how it got there" (Harwell and Gentile, 2000). This approach can include the statistical analyses of relationships between biological response variables (such as benthic invertebrate community metrics) and measured stressors. Various statistical tools are available for these analyses, but the objective is always to partition the variance of observed responses among the various stressors to see which stressors, either alone or in combination, can explain the observed responses (Fairbrother and Bennett, 2000). Integrated response variables (biocriteria, biotic indices, or metrics) are a commonly used tool in observational studies. The relationships between these metrics and stressors have been the subject of numerous publications because of the popularity of their use for the identification of impaired aquatic systems [see Rosenberg and Resh (1993) for a review].

The use of the so-called "reference condition" is one of the current approaches for determining patterns of response to stressors. The reference condition is the range of conditions and biota that exist at reference sites that are considered to be unimpacted or minimally disturbed (Paulson et al., 2001). The reference condition approach involves comparing biological communities at sites exposed to potential stressors with the range of conditions observed at several reference sites. Multivariate bioassessment techniques that use the reference condition approach include the River Invertebrate Prediction and Classification Scheme (RivPacs) and Benthic Assessment of Sediment (BEAST).

The RivPacs project was undertaken in the United Kingdom in the 1980s (Wright et al., 1984, Furse et al., 1984). The RivPacs project had two related objectives: 1) to develop a biological classification for unpolluted running-water sites in Great Britain based on macroinvertebrates, and 2) to determine whether the nature of the macroinvertebrate community at a site may be predicted using physical and chemical features alone. The first goal was intended to provide a classification system for conservation and planning. If the second goal was achieved, it would be possible to judge the severity and nature of degradation at polluted sites from the way the benthic invertebrate community departed from the predicted community.

The RivPacs project was ambitious. Benthic invertebrate samples were collected simultaneously with physical and chemical measurements at multiple sites along every unpolluted watercourse on the island. The final database included data from 267 sites along 41 river systems, each sampled in three seasons. A total of 22 physical variables were measured at each site, along with six standard chemical measurements. The data set was subjected to multivariate analysis using ordination and classification procedures, followed by multiple discriminant analysis to relate the site groups to environmental data (Wright et al., 1984). Later refinements improved the classification and reduced the number of physical variables to eight (Moss et al., 1987).

The RivPacs undertaking was broadly successful, in that a sensible classification of running water sites in Great Britain was obtained. However, because of the large area and broad range of river types covered, gross physical differences, especially geographical location of the rivers (e.g., uplands or lowlands) and the location of the sites along them (distance from source) dominated the classification. Other physical and chemical variables made relatively little difference. Moreover, the predictive capability of the model was imperfect: Wright et al. (1984) reported a success rate of 50-75% at classifying sites into eight or 16 groups defined by the multivariate analysis. Moss et al. (1987) improved this rate slightly. Finally, because the study included only unpolluted sites, it was impossible to judge how polluted or stressed sites would be detected. They may appear as outliers in the classification, or they may be misclassified into a different group.

Despite these shortcomings, the concept of classifying stream sites within an entire region, and thereby detecting disturbance by departures from that classification, is attractive. Rutt et al. (1990) used the same approach on 368 upland streams in Scotland, England, and Wales, and attempted to classify streams according to their susceptibility to acidification. Although a consistent classification emerged, and strong correlations with pH, aluminum and alkalinity were detected, the predictive ability of the scheme was no better than in earlier work.

The RivPacs approach has been emulated in studies in Wales (Ormerod and Edwards, 1987), Australia (Marchant et al., 1985; Marchant et al., 1994), New Zealand (Quinn and Hickey, 1990a,b), and North America (Corkum, 1989). Gradients associated with elevation, channel structure (i.e., gradient and substratum) and dissolved solute concentration have consistently emerged from these studies. Only Quinn and Hickey (1990a) were able to associate specific invertebrate assemblages with disturbance (in particular temperature), enrichment, and degree of catchment conversion to improved pasture, probably because they deliberately included disturbed sites in their 88 rivers. In western Canada and Alaska, Corkum's (1989) classification of 100 sites could distinguish forest from farmland in the surrounding basin.

The BEAST paradigm, originally designed to assess contamination in the Laurentian Great Lakes, supports the development of sediment quality criteria (Reynoldson et al., 1995). Multivariate statistical analysis is performed on benthic invertebrate densities, sediment toxicity, and selected environmental variables at reference sites using cluster analysis and ordination. A model is developed to predict community assemblages and functional responses at new sites with unknown but potential contamination. The predicted community assemblages and functional responses are then compared with the actual benthic communities and responses at the site and the need for remedial action is determined.

As Paulson et al. (2001) point out, the ultimate success of both the RivPac and BEAST methods to detect water quality or habitat impairment is directly related to the degree to which the reference condition characterizations are appropriate. Furthermore, the ability of either of these approaches to produce diagnostic patterns of particular stressors or stressor combinations is questionable.

"Effects-based" monitoring is another category of observational study in which the investigator first identifies whether particular indicators of effects are observed (e.g., fish metrics) and then conducts follow-up studies tailored to confirm the link between the observed effects and potential causes (Munkittrick and McMaster, 2000). The follow-up to effects-based approaches depends on a theoretical framework that attributes particular effects to the actions of particular stressors (e.g., the effects matrices proposed by Gibbons and Munkittrick (1994) for responses by fish populations). The follow-up then proceeds to the appropriate management actions and adaptive management strategies to monitor change.

Two draft reports for WERF projects use the observational approach in the search for patterns of response to multiple stressors. Paulson et al. (2001) have been investigating the ability to discriminate habitat versus chemical limitations using novel data-mining techniques. They also provide a comprehensive review of bioassessment methods and the use of metrics to describe the status or health" of receiving streams. Gerritsen and Jessup (2004) have been investigating the response of stream invertebrates to multiple stressors using data sets from three states.

Paulson et al. (2001) point out that although metrics do a good job of characterizing the differences in aquatic populations, they cannot define what specific chemical or habitat limitations were affecting the aquatic community. They also point out the challenges inherent in identifying and agreeing on reference streams. Unimpacted streams with comparable physical conditions and habitat are hard to find. They emphasize the importance of distinguishing between the effects of habitat degradation and chemical stressors because of the current efforts by the U.S. Environmental Protection Agency (U.S. EPA) and state agencies to implement Total Maximum Daily Loads (TMDLs) for streams. It will be very important to determine how best to spend money on environmental improvements in reductions of chemical loads or in habitat improvements. Paulson et al. (2001) suggest that the traditional reliance on inferential statistics limits our ability to determine complex cause/effect relationships among multiple stressors and the receiving environment. They provide examples of data-mining tools such as Principal Components Analysis (Gauch, 1982), All Possible Regressions (Furnival and Wilson, 1974), and Chi-Square Automatic Interaction Detection (Kass, 1980) to identify key variables in large environmental datasets.

Paulson et al. (2001) demonstrate the utility of the data-mining technique in the search for patterns of response to multiple stressors using four datasets. In the Santa Ana River (California), they found that a combination of habitat characteristics, ammonia and chlorine were responsible for biological impairments in the stream, rather than metals, which were previously believed to be the controlling variable. Thus, the wrong mitigative measures may have been applied had not additional analyses been carried out. They also found that the data-mining techniques worked better on datasets that were not highly diverse; investigation of the Cuyahoga River (Ohio) dataset showed how a wide range of independent variables can make it difficult to detect significant interactions between dependent and independent variables. Nevertheless, some key variables were identified including total number of macroinvertebrates versus cover and water hardness, and relative weight of fish versus dissolved oxygen (DO) and hardness.

The importance of matching habitat data was illustrated in the South Platte River (Colorado). The lack of habitat data made it impossible to determine if there was an impact to biota from habitat changes over time. However, the project team were able to develop a model of the Shannon Weaver diversity score versus orthophosphate and flow. The Fountain Creek (Colorado) dataset was used to validate a user guidance document on the data-mining techniques (Paulson et al., 2001). The model developed from Fountain Creek described invertebrate taxa richness as a function of stream channel embeddedness and orthophosphate concentrations. Examination of the data graphically revealed a linear decrease in taxa richness with increasing embeddedness, whereas orthophosphate had a marginal effect on taxa richness until it reached high levels.

Gerritsen and Jessup (2004) identified and developed predictive, diagnostic relationships that would allow causes of impairment to be identified. Results of the research would form part of the evidence in a SI using EPA's guidance (U.S. EPA, 2000). The project team used three large databases from Florida, Wyoming, and West Virginia in their search for indicators of multiple stressors. Habitat quality was the single most important instream stressor. Metrics that responded to the presence of single stressors always showed an association with habitat. Association of the metrics with multiple stressors was one of degree: the greater the number of stressors, the greater the degradation of the median metric values. The study team was unable to distinguish unique or diagnostic responses of metrics to individual stressor types.

Dyer et al. (2000) conducted a very broad search for patterns in responses. The relationship between fish communities and multiple factors such as habitat, drainage area, gradient, "cumulative effluent," conventional pollutants and chemicals mixtures was explored at the sub-basin, basin, and state level within Ohio. The measures used for fish responses were the index of biotic integrity (IBI), as well as the number of fish species and the percent of gross pathology (e.g., deformities, fin erosions, lesions, and tumors). A geographical information system was used to integrate data; multiple linear regression was used to determine the strength of the stressor-response relationship.

Quality of habitat was generally observed to have a positive influence on IBI whereas high-percent cumulative effluent was a negative influence. Conventional pollutants such as ammonia and biochemical oxygen demand (BOD) were negatively correlated with IBI. Metals and total toxic units were inconsistently related to IBI or the other fish metrics and therefore were interpreted as not being significant drivers of IBI over a large geographic area, although the authors suggested that they may be important at the sub-basin level.

Although these results can be used to generate hypotheses about which stressors may dominate responses in the fish community, they do not provide strong cause/effect evidence. Nor do they provide definitive guidance concerning when mitigation would be necessary. This is because the IBI does not deal in real world units. A reduction of 10 index units at a site relative to a reference site does not necessarily indicate a need for mitigation (Paulson et al., 2001). Furthermore, an IBI could be low because all components are slightly damaged or because one component is severely damaged (Paulson et al., 2001). This incapacity of the IBI to distinguish between general and specific impairment has been cited as one of the major limitations of linear indexes of ecosystem health (Suter, 1993a; Wicklum and Davies, 1995).

The same criticism may be leveled against the rapid assessment methods for biomonitoring (Barbour, 1997) that use a variety of easily measured measurements (metrics) on qualitative samples to deduce the level of impairment in a river relative to reference sites. At best, these methods can only detect whether the site in question differs from the reference sites, and attempt to quantify that difference along an arbitrary scale. They provide no distinction among sources of degradation (e.g., bank erosion versus steel mill effluent), nor do they attempt to differentiate the pattern of response (e.g., loss of a few species or reductions in all species).

The summary indices that are sometimes used in rapid assessment to distil a group of metrics into a single number have been criticized for their arbitrary structure and their lack of interpretive value (Suter, 1993a). Moreover, rapid assessment methods by their nature are insensitive to mild disturbance and therefore are likely to miss any impairment that is not already obvious by other means (Taylor, 1997). If observational methods or effects-based monitoring are to provide any insight into the varying effects of different stressors and their combinations, they will have to steer away from simplistic approaches such as rapid assessment.

Teasing apart the effects of multiple stressors will require a theoretical foundation that will foster predictions about how stressors will interact. Observational evidence is an important part of developing our understanding of multiple-stressor effects; however, it cannot be the sole basis for the assessment of risk. This is because the search for pattern through the use of community metrics, the reference condition, multimetrics such as rapid bioassessment or IBI or effects-based monitoring provides little diagnostic or predictive power. Formal SI methods, although rigorous and based on established and accepted weight-of-evidence approaches, cannot provide a theoretical foundation for the prediction of ecosystem responses to multiple stressors. The U.S. EPA's SI method is an elegant bookkeeping method that assembles our knowledge of the effects of individual stressors, matches that knowledge with the observed responses, and assembles the most plausible explanation for the responses. However, it is retrospective and examines the evidence one stressor at a time. It does not provide a theoretical basis for determining that a particular combination of stressors acting at a particular time and place may have been the cause of the observed response pattern.

Stressor Identification Guidance Document Summary

The SI Guidance Document (U.S. EPA, 2000) outlines the process for linking biological effects with their causes. This process can be complex in aquatic environments where multiple stressors are acting on the system. Therefore, investigation procedures are needed to identify the stressor(s) and appropriate management measures to restore or correct the biological condition of the waterbody.

The SI Guidance Document describes a process to organize and analyze available evidence to determine the cause of biological impairment. The document does not directly address method of collecting the evidence required to do this (e.g., biological assessments, reference condition, data collection, stakeholder involvement). Nor does it identify appropriate management actions.

The SI process is a procedure for analyzing available evidence and determining if the available evidence is adequate to draw a conclusion about the causes of impairment. This process can be applied to any level of biological organization and to any type of waterbody. Similarly, the logic of the process can be applied in single-stressor situations or in complex situations with multiple stressors and cumulative impacts. Although, in more complicated situations, multiple iterations of the process may be required.

The SI process is presented in Figure 2-1. The SI process is initiated by the observation of a biological impairment. Decision-maker and stakeholder involvement is shown along the left-hand side, indicating that their involvement is particularly important in defining the scope of the investigation and listing candidate causes. At any point in the process, a need for additional data may be identified, which is indicated at the right-hand side of the figure. The accurate characterization of the probable cause allows managers to identify appropriate management action to restore or protect biological condition. Management actions are initiated after the stressors have been identified. The effectiveness of the process can be monitored and reiteration of the process can occur if improved conditions are not apparent.

The core of the SI process is presented within the bold line of Figure 2-1. The guidance document focuses on the following three main steps:

- Step 1: List candidate causes of impairment
- Step 2: Analyze data to generate evidence for each candidate cause
- Step 3: Produce a causal relationship using the evidence generated in Step 2 to draw conclusions about the stressors that are most likely to have caused the impairment

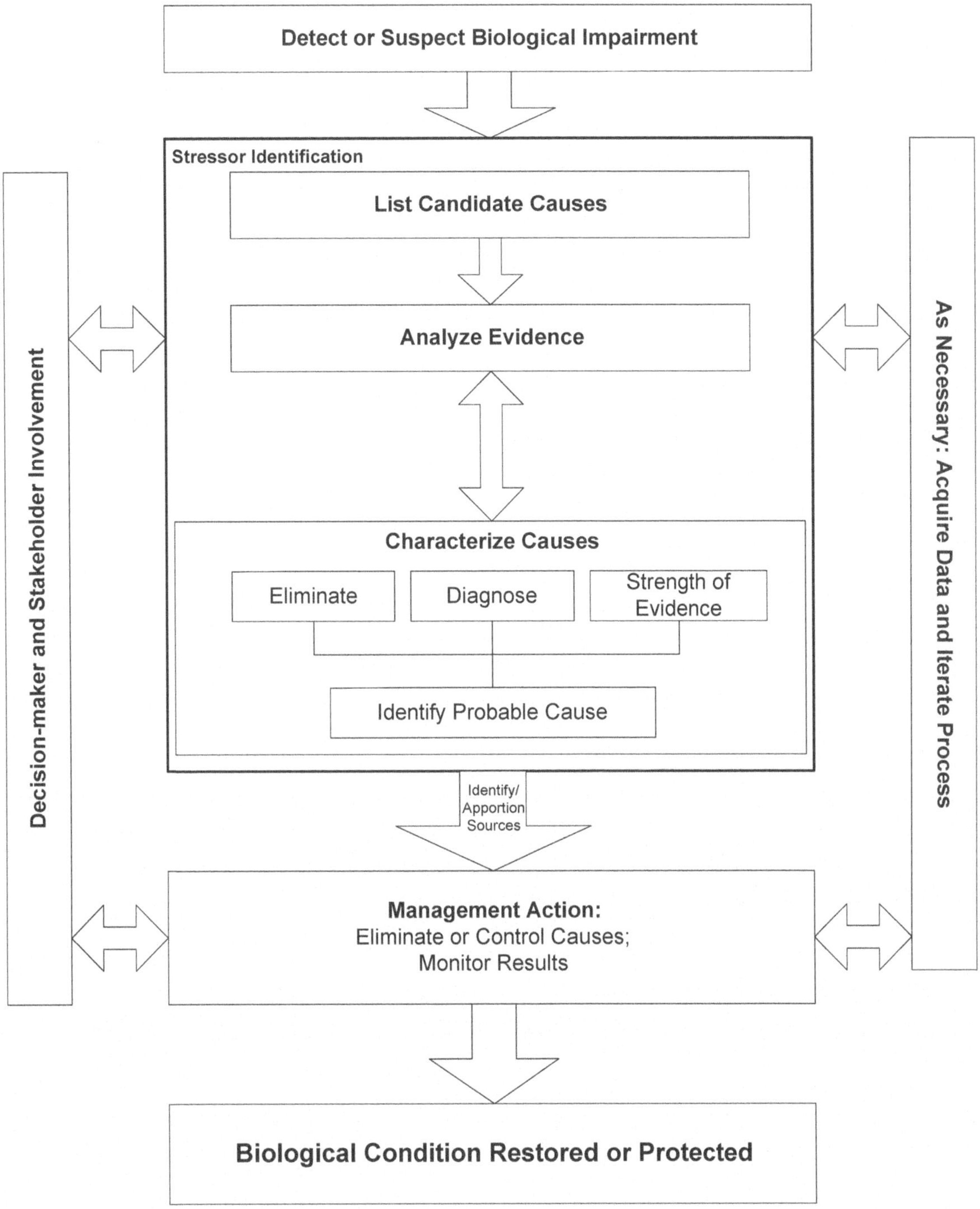

Figure 2-1. Stressor Identification Process.

Step 1: List Candidate Causes. The first step in the SI process is to develop a list of candidate causes or stressors. To do this a detailed description of the effect that prompted the evaluation is necessary. Information on the type of impairment, its magnitude, spatial and temporal extent, and potential causes should be described.

The scope of the investigation is defined in this step. The scope determines the extent and type of data that will be analyzed, the geographic area and time frame under consideration. The scope may be influenced by stakeholders, regulators, cost, logistics, and the extent of knowledge about the impairment. The final part of this step is to develop conceptual models for the candidate causes, linking the causes with the effect. The conceptual model is a pictorial representation of the potential causes of impairment, how these causes may interact, and the mechanisms by which they may cause the effect. The model should indicate where additional data collection may provide useful information. Where multiple causes contribute to cause an effect, the stressor that makes the largest contribution is the principal cause.

Step 2: Analyzing the Evidence. The purpose of Step 2 is to analyze the information that is related to each potential cause identified in Step 1. Potential information may include chemical analyses of water, biota and sediments, biotic surveys and habitat analyses. These may be supplemented with data from laboratory studies or from additional sites; however, the SI process does not require a minimum data set. The data should be analyzed in terms of associations that might support or refute the proposed causal scenarios.

The data in this step are categorized into one of four categories:

1. Associations between measurement of the candidate causes and effects. These are where data are generated from the site/case or from other similar sites/cases. The objectives are to provide evidence that candidate causes and the effect are observed at the same time or place; or when the candidate cause is not observed, the effect is also not observed; or the intensity of the cause is related to the magnitude of the effect.

   The association is more difficult when stressors are dispersed over larger scales, occur intermittently, or cannot be measured. Some stressors (e.g., quickly degraded pesticides, high flow events, or diffuse inputs) may not have measurement data that can be used in specific causal evaluation.

2. Using effects data from elsewhere. The objective is to provide evidence showing that the stressor is present at the study site in sufficient quantity or frequency that data from other sources (e.g., laboratory tests, field tests, etc.) can be used to link the cause with the effect.

3. Measurements associated with the causal mechanism. In developing mechanistic conceptual models depicting the induction of effects, it may be apparent that there are intermediate steps in the causal process that may be observed or measured. Documenting the steps is useful when the end effects of multiple causes are similar, but act through different mechanistic pathways.

4. Associations of effects with mitigation or manipulation of causes. Strong causal evidence can be provided by deliberately eliminating or reducing a candidate cause, and noting whether the effect changes. The manipulations may occur in the field (e.g., placing containers of clean sediment at contaminated locations), or in the laboratory.

Step 3: Characterizing Causes. Characterizing causes involves using the evidence analyzed in Step 2 to reach a conclusion, and to state the levels of confidence in that conclusion. Three methods may be used together in series or independently for characterizing causes. These are:

1. Eliminating alternatives. This is a good option when the set of alternatives is limited, and when disproof does not rely on statistics. However, because of the complexity associated with ecological systems and multiple stressors, many investigations will not have the evidence necessary to confidently eliminate the causes. In addition, it may not be possible to identify a complete set of candidate causes, because one cause may mask another and because the array of possible causes is potentially infinite. Only associations derived from measurements taken from the case under evaluation are strong enough to eliminate an alternative.

2. Diagnostic protocols or keys. If more than one cause remains after the elimination step, diagnostic protocols can be used. Unlike the elimination step, these rely on positive evidence. These are a good alternative when organisms are available for examination, when the candidate causes are familiar enough that they have made it into protocols, and when there is a high degree of specificity in the case, the effect, or both. Examples are certain bill deformities in birds exposed to dioxin-like compounds.

3. Strength of evidence analysis. In many cases, candidate causes are not identified by elimination or diagnosis, and an analysis of the strength of evidence for each candidate cause is required. When there are many candidate causes, or when evidence is ambiguous, strength of evidence analysis can be very useful. Four considerations of strength of evidence that draw on the case itself are co-occurrence, temporality, biological gradient, and complete exposure pathway. Conversely, the considerations of plausibility, specificity, analogy, and predictive performance combine information from the case at hand with experience from other cases. Considerations of consistency and coherency of evidence evaluate the relationships among all of the available lines of evidence.

The relationship between types of evidence and causal considerations is not one-to-one. Each type of evidence may be relevant to several causal considerations, and a causal consideration may be evaluated using several different types of evidence. In many cases, evidence will exist for only some of the causal considerations, and be uneven across candidate causes.

Other methods of causal analysis are available, such as expert judgment. When evidence is ambiguous, the process of expert judgment by a panel may be more acceptable to stakeholders than systematic evidence evaluation.

The results of the causation analysis should be summarized, and the uncertainties concerning that determination presented, and quantified where possible. In most cases, unquantified uncertainties (e.g., incomplete biological data, time of impairment) will dominate. In addition, most causal inferences are based on the strength of evidence, so that no single source of uncertainty characterizes the uncertainty concerning the conclusion. Qualitative judgments are required to characterize uncertainty. In cases where significant uncertainty exists, it will be necessary to determine whether the evidence is sufficient to justify a management action.

The SI process may be iterative, and continue until sufficient confidence in the causal characterization is reached. In the most uncertain and complex cases, the SI process may best serve to guide further data collection, modeling, or analysis efforts.

Iteration Options

Situations where no cause can be identified may result from errors in data collection or analysis, from natural variability of the biological indicators, or from sampling events outside of the appropriate index (e.g., season, biogeographic zone, habitat type, elevation). Reconsideration of the nature of the data and of the impairment may result in additional data collection and causal characterization.

2.3 The Search for Process

An experimental approach is required to search for the processes that govern multiple-stressor responses—including the use of microcosms and mesocosms in laboratory and field settings. The experimental approach requires hypotheses to test. The hypotheses tested are often generated by observational data (i.e., the experiments are part of follow-up studies used to confirm cause/effect relationships). However, the generation of hypotheses should go beyond a case-by-case confirmation of putative cause/effect relationships and extend to generalizations that are useful for prediction.

This literature review focused on the search for studies that present multiple-stressor hypotheses and then test those hypotheses. Because it quickly became apparent that such literature was scarce, the review was then extended to include any experimental studies that involved the use of more than one stressor, even if no hypotheses were presented. It was hoped that the assemblage of experimental studies of multiple stressors would reveal some consistent responses (at least within certain stressor classes). It was also hoped that the experimental studies would indicate whether the commonly used assumption of additivity of multiple-stressor effects was often, seldom, or never valid.

2.4 Risk Assessment as a Framework for the Literature Review

The ecological risk assessment (ERA) method currently in use in the United States and Canada (U.S. EPA, 1992, 1997, 1998; CCME, 1996) is a useful framework for the review of multiple-stressor research. Risk assessment has acquired a reputation for being most applicable to the chemical-by-chemical approach, where, at most, the effects of multiple stressors is approximated by using assumptions of simple additivity and summing a series of hazard quotients. This type of risk assessment seldom extends beyond the toxicological hazards of individual chemicals to individual receptors (with extrapolation to populations and communities).

This reputation has led to some skepticism regarding the applicability of the risk assessment framework to assessments that must take into account the complex interactions that occur in ecosystems. However, the limitations of the risk assessment framework have been in its application, rather than its potential. In our opinion, the risk assessment framework is a flexible and powerful approach that is applicable to multiple stressors. The challenge is to make use of the full range of tools already developed for risk assessments in a manner that truly contributes to our knowledge of the risks from multiple stressors within ecosystems.

One primary advantage of the risk assessment framework is the structure it imposes on the analysis and interpretation of exposure and effects data. The risk assessment framework requires a clear, concise definition of the problem before proceeding to data collection, modeling and analysis. The problem definition stage, when done rigorously, provides a clear conceptual model that generates hypotheses regarding the relationship between stressors and ecological endpoints. The problem definition stage also requires an explicit expression of the ecological assessment endpoints with an accompanying list of measures that can be used to assess whether those assessment endpoints will be met. Thus, decision makers will have results that clearly relate to ecological management goals that integrate ecological and societal values.

The current practice of risk assessment all too often generates vague, nonoperational assessment endpoints (e.g., "sustainable populations of lake trout"). These vague endpoints do not stimulate the development of testable hypotheses. Instead, they lead to risk characterizations that must rely on extrapolations from single-chemical/single-receptor measures to entire populations. These extrapolations are usually highly uncertain and may often be spurious, because the measures may have nothing to do with the variables that actually drive the sustainability of a population. These problems have been recognized, and the development of usable, credible population-level measures and models is an active area of research.

Because of concerns regarding the applicability of the risk assessment framework, the multiple-stressor literature was reviewed for its contribution to conceptual models that present hypothesized relationships between multiple stressors and ecological endpoints. The literature was also reviewed for its use of assessment endpoints, the selected measures of effects, and the relevance of these endpoints and measures to populations, communities, and ecosystems.

Another advantage of the risk assessment framework is the "toolbox" it provides for the assessment of exposure and effects. This toolbox includes standardized single-species and multispecies toxicity tests (including microcosms and mesocosms), standardized biodegradability and bioavailability tests, a wide array of fate and transport models, methods for deriving effects thresholds for wildlife and aquatic life not normally tested in laboratories, field-based effects measures and metrics, a developing suite of population models, and methods for estimating and dealing with uncertainty. These tools can be applied to multiple-stressor assessments. In particular, the most uncertain aspects of multiple-stressor assessment can be identified using the risk assessment framework. These aspects should be the priority items for multiple-stressor research.

The risk assessment framework also provides a bridge between science, the public, and decision makers. The decision-making process required for the management of multiple stressors is not limited to science. Rather, it must also incorporate the public's thresholds of acceptable risk, and the concerns of the public and decision makers regarding long-term sustainability and the consequences of future changes. The risk assessment framework is capable of providing the public and decision makers with a better understanding of the potential for adverse effects on the ecosystem (if the risk communication component of risk assessment is skillfully done).

In summary, the risk assessment framework can help us understand how much risk there is, and how sure we are about that risk. The answers can then be applied to determine the acceptability of the risk.

The following brief description of the risk assessment framework is provided as a guide to how the literature review was conducted. A risk assessment classically has four stages: problem formulation, exposure assessment, effects assessment and risk characterization. Each stage has specific tasks that provide information for the next task. The problem formulation stage defines the objectives of the assessment. The exposure assessment stage determines which components of the ecosystem under study are of concern and the routes and mechanisms by which they might come to harm. Measuring the real changes in the ecosystem and comparing those against expectations are the business of the effects assessment stage. Finally, a quantitative assessment of risk and a decision concerning the advisability of remedial action is made in the risk characterization stage.

2.4.1 Problem Formulation

Problem formulation is a process for generating and evaluating preliminary hypotheses about why ecological effects have occurred, or may occur, from anthropogenic causes. It is the foundation for the entire ERA.

There are three main products from the problem formulation:

- Assessment endpoints that adequately reflect the ecosystem and management goals;
- conceptual models that describe key relationships between stressors and assessment endpoints; and
- a plan for analysis of these relationships (U.S. EPA, 1998).

Assessment endpoints are explicit expressions of the actual environmental value that is to be protected, operationally defined by a receptor (U.S. EPA, 1998). The utility of an assessment endpoint depends on its ability to support risk management decisions. To do this, assessment endpoints must be measurable ecosystem characteristics that adequately represent management goals (U.S. EPA, 1998).

Well-selected assessment endpoints become powerful tools in the risk assessment process. One endpoint that is sensitive to many of the identified stressors, yet responds in different ways to the different stressors, may provide an opportunity to consider the combined effects of multiple stressors while still distinguishing their effects (U.S. EPA, 1998). For example, fish population recruitment may be adversely affected at several life stages, in different habitats, through different ways and by different stressors (U.S. EPA, 1998).

Conceptual models are written descriptions and visual representations of predicted relationships between receptors and the stressors to which they may be exposed (U.S. EPA,

1998). Conceptual models represent relationships between ecosystem processes, receptor responses, exposure scenarios, and land-use activities. They may describe primary, secondary, or tertiary exposure pathways, or they may describe co-occurrence among exposure pathways, ecological effects, and ecological receptors (U.S. EPA, 1998).

Conceptual models have two main parts: 1) a set of risk hypotheses that describe predicted relationships among stressor, exposure and assessment endpoint response (including the rationale for their selection); and 2) a diagram that illustrates the relationships in the risk hypotheses (U.S. EPA, 1998). Risk hypotheses are specific assumptions about potential risk to assessment endpoints and may be based on theory and logic, empirical data, mathematical models, or probabilistic models (U.S. EPA, 1998). Risk hypotheses are formulated using a combination of professional judgment and available information on the ecosystem at risk, potential sources of stressors, stressor characteristics and observed or predicted ecological effects on selected or potential assessment endpoints (U.S. EPA, 1998). Conceptual model diagrams are a visual representation of these risk hypotheses. They are useful tools for communicating important pathways clearly and concisely and can be used to generate new questions about relationships that help formulate plausible risk hypotheses (U.S. EPA, 1998).

Conceptual model development is a source of uncertainty if important relationships are missed or specified incorrectly. If this happens, the risk characterization may misinterpret the actual risks. Uncertainty arises from a lack of knowledge about how the ecosystem functions, failure to identify and interrelate temporal and spatial parameters, omission of stressors, or overlooking secondary effects (U.S. EPA, 1998). In some cases, it may not be known how a stressor is conveyed through the system or affects the receptor.

Multiple stressors are almost always present, and can be the source of confusion, especially when conceptual models focus strictly on single stressors (U.S. EPA, 1998). This general, and rather obvious, statement by the EPA is, in practice, seldom addressed explicitly in risk assessments except perhaps by the use of a general assumption of additivity. Therefore, any contribution to the development of general conceptual models for multiple-stressor situations will be very valuable and was a particular focus of this literature review.

An analysis plan is required in a problem formulation. The analysis plan includes pathways and relationships identified in the problem formulation that will be examined during the analysis phase (U.S. EPA, 1998). The hypotheses that will most likely contribute to risks are targeted. The rationale for selecting and omitting risk hypotheses is incorporated into the plan and includes acknowledgement of data gaps and uncertainties (U.S. EPA, 1998). It may also include an analysis of which analytical approach is best to achieve the goals set out elsewhere in the problem formulation. The analysis plan will likely require revisiting when new data are needed, to address the feasibility of obtaining the new information (U.S. EPA, 1998).

As stated earlier, it quickly became apparent that the analysis plans for assessments of multiple stressors seldom included experimental approaches that are targeted at developing a fundamental understanding of the processes underlying responses to multiple stressors.

Assessment endpoints and conceptual models help risk assessors identify measurable attributes to quantify and predict change. However, the choice of which measures to use when evaluating risk hypotheses is critical to the success of a risk assessment (U.S. EPA, 1998).

There are three types of measures used in a risk assessment (U.S. EPA, 1998):

- Measures of effect—measurable changes in an attribute of an assessment endpoint or its surrogate in response to a stressor;
- Measures of exposure—measures of stressor presence and movement in the environment and their contact or co-occurrence with the assessment endpoint; and
- Measures of ecosystem and receptor characteristics—measures of ecosystem characteristics that influence the behavior and location of the assessment endpoints. Also measures the distribution of a stressor and life-history characteristics of the assessment endpoint that may affect exposure or response to the stressor.

The literature review identifies the measures used in multiple-stressor studies and provides comments on their usefulness and applicability.

2.4.2 Exposure Assessment

Exposure assessment describes potential or actual interaction of stressors with receptors. It is based on measures of exposure and ecosystem/receptor characteristics that are used to analyze stressor sources, location in the environment and the degree of interaction between stressors and receptors (U.S. EPA, 1998). The objective of exposure assessment is to produce a summary exposure profile that identifies the receptor, describes the pathway a stressor takes between the source and the receptor, and describes the intensity and extent of co-occurrence (U.S. EPA, 1998). The profile also describes the impact and uncertainty of exposure estimates and reaches a conclusion about the likelihood that exposure will occur (U.S. EPA, 1998).

Because risk cannot occur unless a receptor is actually exposed to a stressor, the exposure assessment is vital to the development of an understanding of multiple-stressor effects. In some cases, exposure to one stressor may be prevented or greatly reduced by the effects of another stressor. In other cases, exposure to one stressor may increase exposure to another.

2.4.3 Effects Assessment

Ecological effects from exposure to stressors are estimated during effects assessment. The stressor effects are described, linked to assessment endpoints, and evaluated based on how they change with varying stressor levels (U.S. EPA, 1998). Effects data are evaluated to specify the effects that are elicited verify their consistency with assessment endpoints and confirm that the conditions under which they occur are reflected in the conceptual model (U.S. EPA, 1998). The ecological response analysis examines three primary components: the relationship between stressor level and ecological effects, the link between the effects and the stressor, and the links between measurable ecological effects and assessment endpoints when the endpoints cannot be directly measured (U.S. EPA, 1998). Once the effects are identified, an ecological response analysis can be conducted. This will evaluate how the magnitude of the effects changes with the differing stressor levels, as well as confirming that the stressor actually causes the observed effect (U.S. EPA, 1998). The effects can then be linked with the assessment endpoint, and conclusions are provided in a stressor-response profile (U.S. EPA, 1998).

The above guidance provided by the EPA is often extremely difficult to follow, particularly in multiple-stressor situations. Data on varying stressor levels is often lacking. Data relevant to the particular ecological receptor are also often lacking and the assessment must rely on extrapolation from laboratory tests on standard test species. Confirming that the stressor actually causes the observed effects is a very important question that often becomes mired in uncertainty. Experimental evidence that provides exposure-response information for multiple-stressor combinations would be very helpful in reducing some of this uncertainty.

2.4.4 Risk Characterization

Risk characterization is the last phase of ERAs and brings together the planning, problem formulation and analysis of predicted or observed adverse ecological effects related to the assessment endpoints (U.S. EPA, 1998). The risk characterization will allow for clarification of the relationships between stressors, effects and receptors that will provide conclusions regarding the exposure and existing or anticipated effects (U.S. EPA 1998). The results of the analysis phase are used to develop an estimate of the risk to the receptors, as identified in the problem formulation. Risk can then be estimated and put into the context of how significant the adverse effects will be (U.S. EPA, 1998). Then, the uncertainties can be identified and summarized and conclusions can be drawn to aid in decision making.

Risk estimation involves integrating exposure and effects data and evaluating any associated uncertainties. Exposure and stressor-response profiles developed during the analysis plan stage are used in the risk estimation. Risk estimates can be developed using the following techniques: field observational studies, categorical ranking, comparison of single-point exposure and effects estimates, comparisons incorporating the entire stressor-response relationship, incorporation of variability in exposure and/or effects estimates, or process models that use theoretical approximations of exposure and effects (U.S. EPA, 1998).

It is interesting to note that the above list of techniques for developing risk estimates does not include predictive models of multiple-stressor interactions, nor does it include explicit reference to experimental evidence of response to multiple stressors. Hence, risk estimation for multiple stressors has often simply been a compilation of individual risk estimates, stressor-by-stressor, followed by a general and highly uncertain discussion of the potential for additive effects.

The ultimate product of a risk assessment is some version of a "risk narrative" that presents the risk estimate, explains the assumptions underlying the estimate, and describes the level of confidence in the risk estimate.

In complex risk assessments with multiple stressors and/or multiple or very large sites, a weight-of-evidence approach to the estimation of risk is often used. "Weight of evidence is the process by which multiple measurement endpoints are related to an assessment endpoint to evaluate whether significant risk of harm is posed to the environment" (Massachusetts Weight-of-evidence Work Group, 1995). Weight-of-evidence uses explicit criteria for judging the strength of the evidence provided by each measure. "The process of weighing the evidence amounts to determining what estimate of risks is most consistent with the results for all lines of evidence" (Suter, 1996).

The guidance provided in the U.S. EPA Office of Solid Waste and Emergency Response Directive (1999) states that a weight-of-evidence risk characterization requires that sufficient information be collected to make a reasoned decision about 1) causality between levels of contamination and effects; 2) whether the observed adverse effect on the site's local population or community is of sufficient magnitude, severity, aerial extent and duration that they will not be able to recover and/or maintain themselves in a healthy state; and 3) whether these effects appear to exceed the natural changes in the components typical of reference areas.

Lowell et al. (2000) found three steps to be particularly critical in the development of a weight-of-evidence approach for ERA: 1) establishing causality; 2) defining acceptable change; and 3) linking environmental components in a decision-making framework.

Experiments play an essential role in assembling a weight-of-evidence case for the effects of multiple stressors. As Lowell et al. (2000) state, distinguishing among the cumulative impacts of multiple stressors (which sometimes have confounding effects) requires establishing definitive causality links to each stressor. A combination of observational and experimental data is required to support the weight-of-evidence case for the effects of multiple stressors. The strength of causal links could be evaluated by using a formal set of criteria originally developed by epidemiologists, which are outlined in Table 2-1.

Both the search for pattern and the search for process must define significant (or acceptable) change. In most cases, the definition must go beyond a simple statistical determination and extend to an explicit explanation of what a critical effect size would be. This critical effect size is an estimate of the amount of change required to be biologically meaningful —at the population, community, or ecosystem level. The determination of critical effect size requires professional judgment because the links between observations or experimental results and effects on population persistence, community structure, or ecosystem function are highly uncertain.

Legislation and regulation often determine the decision-making framework for the products of a risk assessment. For example, the requirement that TMDLs be specified in waters designated as "impaired" under Section 303(d) of the Clean Water Act has triggered widespread assessment and debate regarding the meaning of "impaired" and the risk management measures that may be required and/or appropriate. Mandated decision frameworks such as the TMDL provisions often create significant constraints in the risk assessment and risk management process. This is particularly true when dealing with multiple-stressor situations, where the theoretical basis for regulation using TMDLs is very uncertain.

Table 2-1. Formalized Set of Causal Criteria Forming Part of a Weight-of Evidence Approach for Ecological Risk Assessment, with Examples of Supporting Evidence from Field Patterns and Experimental Results.

Causal Criterion	Support for Criterion
Spatial correlation of stressor and effect along gradient from more to less exposed areas.	Gradient in effluent/phosphorus concentration downriver from pulp mill/sewage treatment plant positively correlated with algal biomass in river.
Temporal correlation of stressor and effect relative to time course of exposure.	Benthic invertebrate community structure correlated with pulp mill output of effluent solids and phosphorus over 20 years. Algal biomass and chironomid abundance tracked recent upsets/changes in pulp mill effluent treatment.
Plausible mechanism linking stressor and effect.	Increases in growth and abundance, measured in several studies, consistently agreed with known mechanisms for phosphorus/organic matter nutrient enhancement. Negative low DO effects also consistent with expected physiological responses.
Experimental verification of stressor effects under controlled conditions and concordance of experimental results with field data.	In-river nutrient additions and riverside artificial stream experiments verified (and clarified mechanisms for) nutrient enhancement hypothesis, in agreement with field data. Negative low DO effects also experimentally verified.
Strength: steep exposure/response curve.	Artificial stream experiments demonstrated pronounced increases in algal/invertebrate growth with small increases in effluent concentration.
Specificity: effect diagnostic of exposure to a particular stressor.	Increases in growth, measured in several studies, characteristic of nutrient enhancement effects of phosphorus/organic matter additions to oligotrophic rivers.
Evidence of exposure (contaminants or other indicators) in body of affected organisms.	Stable isotopic analyses indicated presence of terrestrially derived (probably pulp mill effluent) carbon and nitrogen in invertebrates and fish downriver from mill.
Consistency of stressor/effect association among different studies within the region being studied.	Nutrient enhancement effects of effluent apparent in all studies conducted, including temporal and spatial field correlations and results from field and streamside experiments. Some toxicity apparent at higher concentrations in two sets of experiments and in field following upsets in effluent treatment.
Coherence with existing knowledge from other regions where the same or analogous stressors and effects have been studied.	Similar nutrient enhancement effects observed in both Thompson and Athabasca River basins. Nutrient enhancement and low DO effects agree with the results of studies conducted in other regions.

Reprinted with permission from Lowell et al., 2000) Causal criteria are derived from Fox (1991); Suter (1993b); and Beyers (1998) and the work of the authors (Lowell et al., 2000); see also Culp, Lowell et al., 2000a. Copyright SETAC, Pensacola, FL, USA.

The ultimate goal for the assessment of risks from multiple stressors is to have an understanding of multiple-stressor effects that is sufficiently robust to support decision making. In other words, we aspire to combine observational and experimental evidence for pattern and process responses to multiple stressors in a manner that allows confident analysis and prediction. The primary outcomes would be credible, defensible, and practical conceptual models that can be used as the starting point for risk assessments at a wide variety of sites.

2.5 Ecosystem Responses to Stress

Because of the important role of professional judgment in risk characterization, it is important that the theory or model used to interpret relative risk is, at the least, referenced and preferably, explained. Several useful models of the responses of ecosystems to stress exist, along with a number of methods for developing a credible risk characterization.

The ecosystem subsidy-stress model of Odum et al. (1979) is one of the best-known models for predicting the response of ecosystems to stressors. Their subsidy-stress perspective holds that the low levels of some stressor can “subsidize” an ecosystem process but that further loading beyond some threshold will “stress” that same ecosystem. Subsidies are assumed to have positive effects on the ecosystem, whereas stresses are negative. An example for aquatic ecosystems would be some anthropogenic nutrient input that may initially increase system productivity, but at higher loading result in undesirable cultural eutrophication. Some stressors, however, may result in ecosystem stress at any input level. Examples of these stressors are toxic materials that eliminate species or inhibit essential processes.

In a subsequent paper, Odum (1985) compiled a list of expected responses in stressed ecosystems at the community and ecosystem levels (Table 2-2). Although Odum’s predictions are based on the oversimplified assumption that undisturbed ecosystems develop along an expected successional pathway, they serve as a valuable tool for developing hypotheses regarding structural and functional responses in disturbed ecosystems. More recent work has confirmed the validity of many of Odum’s predictions for a range of ecosystem types (see Rapport and Whitford, 1999).

A key to Odum’s approach is the assumption that a single major stressor has disrupted the ecosystem from its "nominal" state, thereby producing either a subsidy or stress response. Multiple stressors that act simultaneously are not explicitly considered in this model. For example, an industrial effluent that enters a river can increase water temperature, which can enhance productivity, but also carry a toxic chemical, which kills biota. Therefore, predicted responses can be a complex function of the individual and interacting effects of multiple stressors. For flowing waters that drain varied land use, for example, this condition is probably the rule rather than the exception.

The choice of structural or functional attributes of ecosystems as measures of stress also has been a subject of debate. Several authors (Williams 1980; Woodwell, 1983; Odum, 1985; Schindler, 1987) have argued that structural elements of ecosystems (e.g., community structure, biomass) are more sensitive indicators of anthropogenic stress than are process-oriented functional attributes, because species replacement and other adjustments keep the overall function of the ecosystem steady until the stress is overwhelming. However, Pratt and Cairns (1996) challenge this view and cite examples demonstrating the sensitivity of functional ecosystem parameters. Species-specific approaches also may be most suitable for particular circumstances, such as when threatened or endangered species are the focus. In summary, it is doubtful that a structural or functional attributes alone will be suitable for all cases but rather a pluralistic approach will be needed for evaluating system responses to environmental stressors.

Table 2-2. Trends Expected in Stressed Ecosystems, from Odum (1985).

Ecosystem Property	Expected Trend
Energetics	
1. Community respiration	Increase
2. P/R (production/respiration)	Becomes unbalanced (< or> 1)
3. P/B (production/biomass) and P/B (respiration/biomass)	Increase
4. Importance of auxiliary energy	Increase
5. Exported or unused primary production	Increase
Nutrient Cycling	
6. Nutrient turnover	Increase
7a. Horizontal transport	Increase
7b. Vertical cycling	Decrease
8. Nutrient loss	Increase
Community Structure	
9. Proportion of r-strategists	Increase
10. Size of organisms	Decrease
11. Lifespan of organisms	Decrease
12. Food chains	Shorten
13a. Species diversity	Decrease
13b. Dominance	Increase
General	
14. Inputs/outputs to and from other ecosystems	Increase
15. Autogenic successional trends	Reverse
16. Efficiency of resource use	Decrease
17a. Parasitism and other negative interactions	Increase
17b. Mutualism and other positive interactions	Decrease
18. Functional properties are more robust than structural properties such as species composition	*n/a*

The notion of "ecological resilience" is a pervasive and important underlying concept that affects the development of conceptual models of multiple-stressor effects in ecosystems. Ecological resilience is usually defined as the rate of ecosystem recovery following a perturbation (DeAngelis, 1980; Cottingham and Carpenter, 1994). Resilience has also been described as one component of ecological stability; it is assessed as the rate at which perturbations to stable ecological system decay (Neubert and Caswell, 1996). A broader, and perhaps more useful definition is "the capacity of a system to absorb disturbances" (Raufflet, 2000).

The definition of "recovery" is central to the understanding of the resilience of an ecosystem. The question is "recovery to what state?" The literature can be roughly divided into two points of view: recovery and maintenance of ecosystem function (productivity, mineral cycling) and maintenance of ecosystem structure (species diversity).

The literature emphasizing resiliency of ecosystem function often assumes that recovery will be to a previous equilibrium state where productivity returns to a balance with available resources and mineral cycling is maintained (De Angelis, 1980). Many researchers have observed that functional redundancy is a common characteristic of ecosystems. Biological communities are capable of compensating for the loss of services provided by individual species (Odum, 1985). Lack of resilience in ecosystem function would lead to stress symptoms such as an increased export of unused primary production, nutrient loss and increased community respiration (Odum, 1985). A corollary is that ecosystems are not as sensitive as the most sensitive species because of the redundancy in roles and functions provided by the diversity of species. Such an approach assumes that effects are reversible.

Studies of functional resilience show that resilience increases as the number of cycles that nutrient (or other mineral) atoms make before leaving the system decreases (DeAngelis, 1980). For example, planktivore-dominated (short) food webs are more resilient (as measured by nutrient turnover rate) than piscivore-dominated (long) food webs (Cottingham and Carpenter, 1994). These results would indicate that the short food webs of subarctic ecosystems would have more functional resiliency than longer and more complex food webs present in other geographic areas.

The literature emphasizing resiliency of ecosystem structure stresses the importance of the conservation of biodiversity (Folke et al., 1996; Schindler, 1998). These authors also stress the importance of recognizing that although ecosystems have adapted to natural perturbations such as fire, the response to cumulative effects of stressors such as climate change, increased ultraviolet exposure because of ozone depletion and acidification can produce unexpected and sometimes severe effects (Schindler, 1998). Changes in community structure (species richness, proportion of r-strategists, size of organisms, lifespan of organisms, length of food chains) can affect habitat and food niche availability, as well as functional processes (Odum, 1985).

The “keystone species” concept is related to the concern for the maintenance of biodiversity. Keystone species play vital functional or structural roles in ecosystems; their loss creates a loss of ecosystem resilience because there is insufficient redundancy to compensate. Examples of keystone species are the sawgrass communities of the Florida Everglades (Gunderson, 1999), black gamma grass of the desert grasslands in the American southwest (Whitford, 1995), and the long-lived benthic feeding fish of the Laurentian Great Lakes (Rapport and Whitford, 1999). These species provided vital structural and functional services; their decline in response to multiple stresses such as invasion of exotic species, destabilization of substrates and disruption of the pattern of nutrient cycling has produced major changes in their ecosystems.

Recently, the whole concept of recovery and reversibility has been challenged, at least with respect to the idea that ecosystems can return to a prestressed equilibrium state. These authors contend that ecological communities tend to preserve information about every event in their history and they are not at equilibrium, but rather are in the process of responding to their own unique history of interactions (Matthews et al., 1996; Landis et al., 2000; Sandberg and Landis, 2001). In several microcosm experiments, researchers have “routinely failed” to find recovery as defined as concurrence with the nonstressed state (Sandberg and Landis, 2001). A failure to detect statistically significant differences does not necessarily indicate a return to a prestressed state. This was demonstrated by experiments where functional parameters of systems treated with herbicides were not different statistically but the systems, on re-exposure, were much more resistant to the herbicide (Kersting and Van Wijngaarden, 1999). These results have produced the theory of "community conditioning." This theory suggests that ecological systems are complex, nonequilibrium systems and that our assumption that ecosystems can be "managed" to preserve or return to some prestressed state is spurious (Landis and McLaughlin, 2000). Instead, ecosystem management must recognize that the postdevelopment state of the ecosystem will be different—it may be equally functional with the desired structure, but it will not be the same as before development.

The debate regarding the nature of resilience and recovery leads to the need for clear definitions of the desired state of the ecosystem. "To decide that certain structural changes will be allowed as long as perceived function is not affected is a regulatory decision, not an ecological one" (Landis et al., 2000). Further, reliance on a 'no observable effect' strategy is at odds with ecological reality; a natural community is too complex to be reducible to an index and too unique to provide an unequivocal standard for comparison (Matthews et al., 1996). "All systems change, all the time, and our actions are an inevitable part of that change" (Matthews et al., 1996).

Landis and colleagues (Landis and Wiegers, 1997; Wiegers et al., 1998; Obery and Landis, 2002) have continued to develop their alternative view of the properties and response of ecosystems. Landis (2002) presents the following thought-provoking points:

- Complex structures are neither completely deterministic nor completely stochastic. A conceptual model for describing an ecosystem should incorporate both deterministic and stochastic inputs. Depending on the initial state, a wide variety of endpoints could result from the same initial conditions depending on the sensitivity of the system to the stochastic variables.
- The causes and effects of the events experiences are not proportional. Organism-level toxicology is based on the principle of dose/effect relationships. Ecosystems may not respond this way. Step functions or even a reversal of effect may occur, depending on points 3 and 4 below.
- The different parts of complex systems are linked and affect one another. The type and number of ecosystem components is constantly changing. Therefore, the type and magnitude of interactions and effects will change.
- Complex systems undergo irreversible processes. Returning to an original state is not a property of complex systems.
- Complex systems are dynamic and not in equilibrium; they are constantly moving targets.

The target state can change even as the measurements take place and no ground or nominal state exists for the system. The measurements that would be most useful in predicting future events are the position of the system in ecological space and the trajectory. However, these properties are difficult to measure.

Landis (2002) states that "the failure to incorporate both stochastic and deterministic features, the assumption of proportionality between dose and effect, the search for recovery, and assumptions of independence of effects all will lead to an increase in uncertainty." This statement is followed by some advice on how to reduce uncertainty. The advice is as follows:

- Because reference or control sites do not exist, use gradient or other approaches that specifically deal with environmental heterogeneity. A catalogue of the history of the site under investigation may help predict future responses to stress. Do not expect the same outcome every time the system is stressed.

- Recognize that the strength and pattern of responses will be linked to landscape features. Spatial factors affect the types and frequency of outcomes. Localized contamination can impact populations at a distance if they are connected by immigration. Consider whether the population being stressed is a source or a sink population. Consider also the distance among patches because this controls the rate of migration and the degree of interaction. Relative arrangement between patches in regards to migration patterns governs the sequence of events leading to extinction or persistence of a population. Initial population size within a patch is an important input as well as the potential total population possible in each. Population declines or increases at a site may be caused by disturbances at other sites, and at a time well in the past.

- Communicate to the decision makers and stakeholders the importance of the above. Explain that a reference site approach will lead to uncertainty in the decision-making process. Demand that data be obtained to place the species serving as assessment endpoints in a spatial context. Put in context the importance of understanding the historical events that provide clues to the past population dynamics and spatial context of the site (Landis, 2002).

Although portions of the above advice would gain ready acceptance (particularly the advice regarding the importance of spatial context), the assertions that there is no such thing as a reference site and that ecosystems do not "recover" are controversial. Although these assertions are intriguing and have an experimental and theoretical basis, they tend to cause a certain amount of paralysis when one is faced with the requirement to make decisions about mitigation or remediation measures.

How does one assess whether a certain amount of "impairment" requires remediation if we don't know where the ecosystem may be heading? How do we measure our success in remediation against a back drop of constant change along a unique trajectory? Somehow, we still need some sort of measure of the "state of the ecosystem" versus some idea of the desired state. Is there not some middle ground between the rigid adherence to Clementian ideas of ecosystem recovery and the appealing, but difficult-to-implement, ideas about community conditioning? The authors of this review believe that there is a middle ground. However, the definition of that middle ground will be difficult or impossible unless we make progress in our fundamental understanding of how particular combinations of stressors may be expected to behave and how particular ecosystems may be expected to respond. Hence, the central focus of the literature review was on studies that contributed to our understanding of the relationship between particular combinations of multiple stressors and ecosystem response.

CHAPTER 3.0

LITERATURE COMPILATION, REVIEW, AND ANNOTATED BIBLIOGRAPHY

3.1 Methods

To explore the literature on multiple stressors, a simultaneous search was conducted of over 30 databases, including all of the major sources for published work and government literature (Appendix A). The literature search produced a list of approximately 600 titles. Therefore, the first step in refining the title list was to scan all the titles and delete those that were clearly irrelevant.

The screening of the titles left 386 references as candidates for inclusion in the final database. These titles were distributed amongst the team members for review. This served a twofold purpose: to determine which papers were sufficiently relevant and useful to be included in the database, and to provide annotation, including a quality ranking for the papers in question.

Approximately 50 of the candidate titles from the literature search turned out to be irretrievable. These were all conference abstracts or unpublished reports. Of the 336 reviewed papers, 178, or approximately 53%, were rejected as unsuitable by the reviewers. New papers have since been added.

Consequently the database currently contains 242 relevant, good quality articles. The small number of acceptable papers clearly indicates that the effect of multiple stressors is a relatively new area of research. All of the relevant articles have been compiled into an annotated database that is fully searchable by any of the keyword fields and is available by request.

3.2 Quality Control

As a result of the variety of literature types, sources reviewed and the large number of reviewers involved, there is a possibility of subjectivity in the selection of acceptable papers. As a quality control check, the quality control officer for the project reviewed a large, random subset of the entire database (including both accepted and rejected papers).

There was no marked difference in rejection rate amongst reviewers. There was a slight disparity in the rate of rejection between the participants who reviewed conference proceedings and abstracts and those who dealt with journal papers, but the difference was not large. An inspection of abstracts for rejected papers (where available) and reasons given for rejection revealed no apparent biases or differences among reviewers, and in almost all cases the quality control officer agreed with the original reviewer. Unacceptable papers are usually obvious. Hence, quality of the review process is acceptable.

CHAPTER 4.0

REVIEW OF THE MULTIPLE STRESSOR LITERATURE

4.1 The Application of the Risk Framework

The review of the multiple-stressor literature first examined how well the research problem was defined. Problem formulation requires:

- identification of the stressors of concern;
- identification of the receptors of concern (including the level of ecological organization - individual, population, community or ecosystem);
- description of the routes (pathways) of exposure or other descriptors of the conditions that describe the severity, duration and extent of exposure;
- assessment endpoints (the ecological endpoint(s) of concern); and
- measures (the indicators of exposure and effects).

In addition to these basic requirements, it was desirable to know whether the study addressed a current or future potential problem (i.e., retrospective or prospective study). The rationale for selection of the particular stressors, receptors and endpoints was also desirable.

The literature on multiple stressors is deceptive in that many papers purporting to study combined or cumulative effects from different sources do not include the critical information or experiments. A great many "multiple-stressor" papers merely list the various stresses to which an ecosystem or watershed is subjected and comment on the likely effects of each (Bailey et al., 1996; Boët et al., 1999). Other papers study each stressor separately without considering combined effects (e.g., Hatakeyama, 1989; Kinross et al., 2000) or report on the effects of mixtures without sorting out which stressor is responsible for which response.

Problem formulation in a number of the papers conformed, to a greater or lesser extent, to the problem formulation step of the risk assessment paradigm. The extent to which they did so was often a function of the type of study that was being conducted. Many of the papers made reference to the need for multiple-stressor studies in preference to single-stressor tests in addressing questions involving ERA (Marshall et al., 1981; Clements, et al., 1989; Forbes, 1991; Hanazato, 1991; Genter, 1995; Horne and Dunson, 1995; Carder and Hoagland, 1998; Growns et al., 1998; Jak et al., 1998; Culp et al., 2000; Stuijfzand et al., 2000).

Single-species, multiple-stressor studies (Hartwell et al., 1987; Forbes, 1991; Lowell et al., 1995a; Schlueter et al., 1997) that investigated a relatively small segment of a more complex ecological problem (identified or not) tended not to be explicit in identifying any risk assessment goals in their studies. They would generally address things like identification of stressors, receptors, exposure pathways and measures, but often made no explicit mention of assessment endpoints within an ERA-type context.

Larger, more complex studies like the multispecies mesocosm experiments (e.g., Hoagland et al., 1993; Horne and Dunson, 1995; Lowell et al., 2000; Culp, Lowell et al., 2000a,b) often employed some form of weight-of-evidence approach. The essential requirement of ecosystem mimics like microcosms and mesocosms was implicit and these studies expressly recognized the limited ability of observational studies to do much beyond describe the pattern of receptor responses and recognized the need for experimental work to address questions of mechanism and to provide information with the requisite causal conclusions to be of predictive value.

A large-scale multiple-stressor study is being conducted by the U.S. EPA's Office of Research and Development (Carpenter and Lunetta, 2000). The study covers an area of approximately 420,000 km^2 that includes part or all of eight states and the District of Columbia and is called the mid-Atlantic study area. The study exemplifies the use of a risk assessment framework to manage large projects, including problem formulation, SI and determination of receptor endpoints for the exposure assessment (Carpenter and Lunetta, 2000).

This review includes a discussion of the choices of exposure pathways in terms of the rationale provided by the authors, if any. We examine the range of concentrations and/or doses used and compare these with what would be environmentally relevant. The review includes examination of the measures used for each of the four main stressor groups. The relevance of the measures to our understanding of ecosystem response to multiple stressors is discussed. The overall contribution of the literature to the development of conceptual models of multiple-stressor effects on aquatic ecosystems is critically examined.

4.2 Studies of Metal Combinations and Metals Plus Other Stressors

Studies of combinations of metals are the most common multiple-stressor studies in the literature. The metal combinations, receptors and measures used in these studies are summarized in Table 4-1.

Table 4-1. Stressors, Receptors, and Measures Used in Studies of Metal Combinations.

Stressors	Receptors	Measures	Reference
Fish Receptor Studies			
Arsenic, cadmium, copper, lead and selenium	*Morone saxatalis* juveniles (striped bass)	Infection with *flexibacter comumnaris*	MacFarlane et al., 1986
Nickel, chromium and cadmium	*Oncorhynchus mykiss* (rainbow trout)	Morphometric measurements of the structure of the gill secondary lamellae	Hughes et al., 1979
Arsenic, cadmium, copper, lead and zinc	*Oncorhynchus mykiss* (rainbow trout)	Survival, growth, metal uptake	Mount et al., 1994
Arsenic, cadmium, copper, lead and zinc (in field-collected invertebrates used as food)	*Oncorhynchus mykiss* (rainbow trout)	Metal accumulation in tissues, lipid peroxidation of kidney, gross abnormalities including scale loss, ion regulation	Farag et al., 1994
Copper, chromium, arsenic, selenium	Fathead minnow	Avoidance	Hartwell et al., 1987

Table 4-1. Stressors, Receptors and Measures Used in Studies of Metal Combinations (continued).

Stressors	Receptors	Measures	Reference
Copper and zinc—this appears to be a single-stressor study, i.e. the effects of the two metals were examined in separate experiments	*Epeorus-latifolium* (mayfly)	Growth, emergence	Hatakeyama, 1989
Aluminum, pH, 2 levels of Ca^{++}	*Salvelinus fontinalis*	Whole body electrolytes	Wood et al., 1988a
Aluminum, pH, 2 levels of Ca^{++}	*Salvelinus fontinalis* (brook trout)	Sodium losses associated with mortality, metals deductions in tissue (liver and plasma)	Booth et al., 1988
Cobalt, copper	*Oncorhynchus mykiss* (Rainbow trout)	Acute toxicity, survival	Marr et al., 1998
Cobalt, copper	*Oncorhynchus tshawytscha* (Chinook Salmon) and *Oncorhynchus mykiss* (Rainbow trout)	Neurobehavioral responses and behavioral avoidance	Hansen at al., 1998
Cadmium, copper with either neutral pH or pH of 4.8 in the presence of calcium, sodium, potassium and chloride	*Salmo gairdneri* (Rainbow trout)	Effect on Ca influx, net Na flux in both juveniles and adults. Effect of metals on Na influx in relation to pH and Ca. mortality	Reid and McDonald, 1988
6 metals (Al, Fe, Zn, Mn, Ni & Cd, or Pb)	*Salvelinus fontinalis* (brook charr)	Body ion loss, Biomarker for measuring impact on reconstituted coal mine water	Grippo and Dunson, 1996
6 metals (Al, Fe, Zn, Mn, Ni, Cd)+ pH, sulphate, conductivity, hardness	*Salvelinus fontinalis* (brook charr)	Body ion loss, Biomarker. Na and Ca flux	Grippo and Dunson, 1996
Aluminum, cadmium, chromium, copper and zinc	Striped bass	Toxic units, risk estimation , survival	Logan and Wilson, 1995
Sulfate, selenate	*Artemia* sp. (brine shrimp)	96 h LC50, developmental assays on emergence and hatching	Forsythe and Klaine, 1994
Aluminum, manganese, iron, nickel, copper, lead, and zinc in the presence of soft and acidified water	*Jordanella floridae* (American flag fish)	Reproduction, survival, size, life cycle stages	Hutchinson and Sprague, 1986
Cadmium, copper and zinc	*Oncorhynchus tshawytscha* (juvenile Chinook salmon)	96 h LC50	Finlayson and Verrue, 1982
Cadmium, copper and zinc	*Catostomus commersoni* (White suckers)	Populations study, growth parameters, fecundity, maturation	McFarlane and Franzin, 1980
Cadmium and zinc	*Jordanella floridae* (American flag fish)	96 h LC50, survival, growth, reproduction	Spehar, 1976
Algae Receptor Studies			
Cadmium and zinc	zooplankton	Community assemblage	Marshall, Mellinger, and Parker, 1981

Stressors	Receptors	Measures	Reference
Aluminum and copper, in presence of sodium, potassium, calcium, or magnesium	*Chlorella saccarophila* (green alga)	Growth	Folsom et al., 986
Cadmium, cobalt, copper	*Chlorella vulgaris* (green algae)	Growth response, 96 h static EC50	Rachlin and Grosso, 1993
Silicon, nitrate and phosphate	Diatom species	Growth, algal chlorophyll, biovolume, taxonomic shift in population	Carrick and Lowe, 1988
Aquatic Fungi Receptor Studies			
Cadmium, copper, and zinc	Aquatic hyphomycetes	Growth, synthesis of sulphur-rich compounds, Glutathione levels	Miersch et al., 1997 (in German)
Daphnia Receptor Studies			
Arsenic, molybdenum and selenium	*Ceriodaphnia dubia*	Sublethal interactive effects of chronic exposures. Survival & fecundity	Naddy et al., 1995
sulfate, selenate	*Daphnia magna* and *Chironomus decorus*	Acute 48-h lab protocols, bioaccumulation	Hansen et al., 1993
Chironomids Receptor Studies			
Zinc in the presence of 5 different Salinities	*Chironomus riparius*	Effect of salinity on the uptake of zinc. Zn uptake depends on ion activity rather than concentration	Bervoets et al., 1996
Mussel Receptor Studies			
Cadmium, copper, zinc, lead and nickel	*Pyganodon grandis* (fresh water mussel)	effects of metal mix on Cd bioavailability and uptake	Stewart, 1999
Aquatic Macrophyte Receptor Studies			
Zinc, copper, cobalt, and chromium	*Lemna minor* (duck weed)	Toxicity, antagonistic, synergistic or additive effects	Ince et al., 1999
Iron, aluminum, copper, zinc, silicon, cobalt, chromium, nickel, lead, cadmium and strontium	8 different macrophytes	Uptake rate, species specific adsorption, tissue specificity	Bosserman, 1985
Both major and trace elements from coal ash drainage water (22 elements) in the presence of abiotic factors like settling	*Lemna perpusilla* (duck weed)	Accumulation of toxic metals	Rodgers et al., 1978

Table 4-2. Stressors, Receptors, and Measures Used in Studies of Metals Plus Other Stressors.

Stressors	Receptors	Measures	Reference
Chromium plus a fungal parasite (Saprolegniales)	Carp	Egg mortality and fungal growth	Draggan, 1977
Cadmium and predation	*Pimephales promelas* (Fathead minnow) (prey) and *Micropterus salmoides* (largemouth bass) (predator)	Prey vulnerability, Abnormal schooling behavior	Sullivan et al., 1978
Titanium and temperature	*Oncorhynchus mykiss* (rainbow trout)	Hematocrit; plasma sodium, chloride, glucose, and lactate; erythrocytic ALA-D activity	Lehtinen et al., 1984
Copper, prior population stress	Fathead minnows	Survival	Schlueter et al., 1997
"Metals" and particulate matter	*Chironomus riparius*	Larval growth	Stuijfzand et al., 2000
Salinity and cadmium	*Hydrobia ventrosa*	Growth	Forbes, 1991
Aluminum and pH	Freshwater benthos	Population and community metrics	Genter and Amyot, 1994
Cadmium and pH	*Leptophlebia-marginata* and *Baetis-rhodani*	Cadmium uptake, emergence, survival and locomotory activity	Gerhardt, 1990
Municipal wastewater (contained "organic enrichment and metals (cadmium, copper, iron and zinc); toxicity tests using cadmium, copper and humic acids	*Dreissena polymorpha* (zebra mussel) and *Chironomus riparius*	Filtration rates of zebra mussels; growth in midges	Stuijfzand et al., 1999
Aluminum, low pH	Periphyton	Algal abundances	Genter, 1995
Aluminum, low pH	*Salvelinus fontinalis* (brook trout) and *Salmo salar* (Atlantic salmon)	Mortality, growth, swimming activity and gill morphology	Smith and Haines, 1995
Low pH, aluminum, copper, iron, lead and zinc and tannic and humic acids	Pond community three amphibian species, one mosquito genus and 18 genera of algae	Time to metamorphosis, survival, wet body mass at metamorphosis (amphibians). Mosquito emergence. Algal community assemblage	Horne and Dunson, 1995
Zinc and pH	Periphyton		Dean-Ross, 1989
Copper and diazinon	*Ephoron virgo* (mayfly) larvae	Mortality	Van Der Geest et al., 2000
Methyl mercury and thiolate	*Cyprinodon variegatus* (sheepshead minnows)	Body burden	Leaner and Mason, 2001
Copper and water quality (alkalinity and hardness)	Macroinvertebrate community	Species abundance and diversity	Clements et al., 1989

Stressors	Receptors	Measures	Reference
Aluminum, low pH, organic matter	Stream invertebrates (*Asellus intermedius*, *Pycnopsyche guttifer*, *Lepidostoma liba*, *Nemoura* sp., and *Physella heterostropha*)	Mortality	Burton and Allan, 1986
Zinc and pH	Bacteria	Abundance and activity	Dean-Ross, 1991
Arsenic and PCB	*Chlamydomonas*	Mortality, toxic threshold	Christensen and Zielski, 1980
Heavy metals (zinc, aluminum, iron, etc.), pH	*Ulothrix*	Biomass	Niyogi et al., 1999

4.2.1 A Brief Synopsis of Studies of Metal Combinations

A large body of literature exists on the interaction between metals and physical variables, particularly pH and hardness. For example, Gerhardt (1993) reviewed how heavy metals and acidity can interact to affect stream invertebrate communities. For the most part however, these physical variables are modifying factors that influence the toxicity of the metals but are not themselves stressful. Therefore, these studies were not included in this review. The only exception would be pH, which can be a stressor at the low end of the environmental range. Low pH coupled with toxic metals routinely occurs only in two situations: 1) in acid mine drainage or similar runoff from acidic, metal-bearing rocks; and 2) in "acidified" soft waters in which the stress from low pH is often exacerbated by monomeric aluminum rendered soluble by the acidity (Baker and Schofield, 1982; Hall et al., 1987). These two sources of combined stress are briefly considered as part of the discussion of metals in combination with other stressors.

Niyogi et al. (1999) examined the effects of pH, dissolved metals (primarily zinc, aluminum, and iron), and metal hydroxide deposition on the biomass of the alga *Ulothrix*. They concluded that metal hydroxide deposition of aluminum and iron was controlled by pH and that biomass of *Ulothrix* was negatively affected by the deposition of aluminum hydroxides but not iron hydroxides. Gerhardt (1993; p. 309–310) provides a short review of metal interactions but acknowledges that studies of metal interactions are rare. Scarce literature exists on metals and organics combinations, with the notable exception of Van Der Geest et al. (2000).

Horne and Dunson (1995) examined the effects of multiple metals in combination with low pH and dissolved organic carbon (DOC) on a palustrine assemblage of three amphibian species, one mosquito genus, and 18 genera of green algae. They found that community structure at one trophic level (here, amphibians) could be affected by the individual or the combined effects of potentially toxic substances either directly, through effects on mortality, growth rates and fecundity, or indirectly, through effects on lower trophic levels (here, algae) on which they feed.

Numerous studies of acid lakes and acid mine drainage have demonstrated that the effects of pH and metals are very difficult to disentangle. Eggs and larvae are generally most susceptible to acute acidity and aluminum, but chronic exposure to low pH may be equally debilitating to adult fish. McDonald and Milligan (1988) showed that 10-week exposure of adult brook trout to pH 5.2 caused an ultimately fatal loss of Na^+ across the gill membrane. The gill did not repair when pH was raised to 6. However, exposure of fish to low pH in the presence of aluminum induced a compensatory mechanism that more than overcame the Na+ loss. Research on benthic invertebrates has shown stimulation of behavioral drift by exposure to combined low pH and aluminum in some studies (Hall et al., 1987), whereas in others drift or mortality was largely attributable to pH alone (Allard and Moreau, 1986, 1987). Crayfish are very sensitive to reductions in pH and calcium ion concentrations, but dissolved aluminum contributes little additional toxicity (Berrill et al., 1985).

A large amount of research has gone into the lake acidification effects on aquatic ecosystems in Canada (National Research Council canada 1981; Environment Canada, 1997). The overall goal of this project was to determine cause/effect relationships between surface water acidification and fish population responses by using bioassay data from the laboratory together with water quality and fish population data from the field. The physiological responses of freshwater fish to environmental acidity have been studied extensively over the past 25 years (Brown, 1983; Freeman and Everhart, 1971; Campbell and Stokes, 1985; McDonald, 1983a,b; Neville, 1985; Wood et al., 1988a,b,c). The response in adult brook trout survival, growth and fecundity to long-term exposure to acid aluminum and low calcium levels was studied by Mount et al. (1988). Vitellogenesis and osmoregulation under the same exposure conditions were also studied (Mount et al., 1988).

Various chronic effects were studied in brook trout affected by acidic conditions and aluminum. These effects included ion regulatory and respiratory disturbances (Wood et al., 1988a), blood composition, and net sodium fluxes (Wood et al., 1988b,c). Studies were also conducted to determine the chronic effects of aluminum and low pH (McDonald and Milligan, 1988), the effects of these stressors on ventilation (Walker, Wood and Bergman, 1988) and the effects of these stressors on morphometric changes in gill secondary lamellae (Tietge et al., 1988).

These examples serve to illustrate the difficulty of quantifying the severity of multiple stressors that interact in a nonlinear manner. The severity of an aluminum stress to aquatic organisms depends on the pH, but a monotonic increase in aluminum concentration or decrease in pH will not necessarily produce a uniform increase in the stress to aquatic life. With the benefit of extensive research on this interaction, however, we can produce reasonable predictions of cumulative stress either from chemical equilibria predicting the proportion of $Al(OH)^{2+}$ or operationally from the observed effects on aquatic organisms. It would be a greater challenge to calculate the cumulative stress from two stressors whose interaction has been less completely worked out.

Clements et al., (1989) examined changes in the predation risk of two species of trichopterans by a stonefly as a function of copper exposure. They found that the risk of predation increased in both species in the presence of copper. This contradicts Peckarsky (1983) and others who proposed that predator-prey interactions would become less significant with increased environmental stress. This suggests if predation pressure in such systems increases as a function of stress, the structure of benthic communities may be indirectly affected.

Because of the paucity of information on the combined effects of acidification and heavy metal toxicity on invertebrates, Gerhardt (1990) examined the effects of subacute levels of cadmium at two pH levels on uptake, emergence, locomotory activity and survival of two species of mayflies. He found that different combinations of the cadmium/pH stressors elicited species-specific responses for each of the endpoints measured. These differing responses were thought to be due, in part, to the individual life histories of the species under study (i.e., one species lives in lakes and slow-moving water whereas the other is resident in the high-velocity portions of rivers and small streams). The paper limited its discussion of multiple stressors to their effects on the two invertebrates examined, and made no attempt to place the findings within a broader risk assessment framework.

Stuijfzand et al. (1999) examined the responses of a chironomid to the combined effects of copper and organic material/humic acids. They found that humic acids (typically not present in standard metal toxicity tests) mitigated the toxic effects of copper on chironomids. In another paper, Stuijfzand et al. (2000) reported that growth in the presence of both a metal and organic matter (e.g., humic acid) was greater than in a nonmetal, high-food control. From this a mechanism was speculated that although toxicants induced negative effects, these effects were obscured by the superior nutritive quality of the particulate organic matter. Similar results were reported by Lowell et al. (1995a) where it was found that effluent-exposed mayflies grew faster than even high-food control animals. One possible explanation for this response was the increased nutritive value of food in the effluent treatments in comparison with the high-food controls (Lowell et al., 1995a).

Marshall et al. (1981) studied the effects of separate and combined additions of cadmium and zinc on a zooplankton community. They concluded that the potential combined effects of increased concentrations of cadmium and zinc on zooplankton communities would be primarily caused by zinc because it reduces cadmium adsorption by plankton.

Stuijfzand et al. (2000) evaluated the interactive effects of toxicants (either complex mixtures or metals) and organic matter in river water on macrofauna through manipulations of both factors. They found that particulate (organic) matter had the potential to reduce the inhibiting effects of toxicants on the study species not only by reducing bioavailability of the toxicants but by serving as a supplementary, superior food source.

Forbes (1991) evaluated the effects of cadmium exposure and fluctuating salinities on the growth of snails. Forbes concluded that the altered physiological state induced by increasing salinity rendered the snail less able to compensate for increased metabolic demands incurred during cadmium exposure.

Genter (1995) investigated whether aluminum or acidity was responsible for changes in community composition of benthic algae. Genter surmised that the effects of the stressors alone or in concert were species-specific thus demonstrating the utility of algal community composition for indicating acid and aluminum stress.

4.2.2 Review in the Context of Risk Assessment

In most cases, the receptors in studies of metal combinations were fish or invertebrates, especially benthic insects (Tables 4-1 and 4-2). However, algal studies were also represented

(Folsom et al., 1986; Christensen and Zielski, 1980; Dean-Ross, 1989; Genter, 1995; Horne and Dunson, 1995; Niyogi et al., 1999).

Many of the studies in the previous tables examined benthic invertebrates and the effects on these communities by metal and other stressors. Benthic invertebrates have been used in certain metal combination studies to investigate questions such as general benthos tolerance (Burton and Allan, 1986; Genter and Amyot, 1994; Van Der Geest, 2000) and species-specific tolerance (Clements et al., 1989; Gerhardt, 1990; Stuijfzand et al., 1999, 2000). In general, the response of a species to multiple stressors differs substantially from its response to the presence of a lone stressor.

Some studies provided explanations for the selection of the test species. For example, the decline of striped bass populations along the middle Atlantic coast of the United States was the impetus for the selection of this species in a study of the effects of five metals on susceptibility to bacterial infection (MacFarlane et al., 1986). The interaction between metals and parasite infections was of interest in a lake serving as a major nursery area for sockeye salmon (Boyce and Yamada, 1977). In many cases, standard test species were used (e.g., rainbow trout, fathead minnow). Sometimes, a combination of a standard test species and a nonstandard species was used (e.g., fathead minnow and largemouth bass (Sullivan et al., 1978)). There was seldom any explanation of whether the standard test species were appropriate surrogates for wild fish that may have been exposed to the particular metal combination in the field. Authors who did include the suitability of test species included Leaner and Mason (2001). Leaner and Mason (2001) chose sheepshead minnows (*Cyprinodon variegatus*) as their test species because it was a U.S. EPA-approved species commonly found in pelagic environments of the Chesapeake Bay.

The routes of exposure in studies of metal combinations were usually via water (MacFarlane et al., 1986; Draggan, 1977; Hughes et al., 1979). However, a few studies included exposure via the diet (Farag et al., 1994; Mount et al., 1994; Woodward et al., 1994).

The exposure of receptors to metal mixtures was almost always expressed as concentrations in water. Occasionally, exposures expressed as metal concentrations in sediments (e.g., metal hydroxide deposition in g $m^{-2}d^{-1}$; Niyogi et al., 1999) and food (Mount et al., 1994; Farag et al., 1994; Woodward et al., 1994) were used. In rare cases, the concentrations in media were related to actual uptake of metals into tissues (Mount et al., 1994; Farag et al., 1994).

Some studies of metal mixtures included measurement of tissue concentrations (Golder, 2001). In these studies, the effects data can be related to actual metal uptake. However, in most cases, exposure is described by concentrations in water or sediments. This creates problems when interpreting the relationship between the "exposure" and effects because the actual metal dose (metal taken up into tissue) is affected by a number of environmental variables that control metal speciation and bioavailability. These variables include:

- Water pH, hardness, total suspended solids, and temperature;
- Sediment texture, total organic carbon content, acid volatile sulfide content, iron/manganese oxide concentrations, and redox potential (Chapman et al., 1998); and
- Receptor life stage and ecological niche.

If the study cannot account for the effect of these variables on actual dose received by the receptors, the statistical relationships reported between metal concentrations and effects may be spurious.

The ecological assessment endpoint, as defined within risk assessments, was rarely, if ever, stated. The objectives of most studies were usually particular research questions (e.g., whether essential metals and nonessential metals produced different effects on gill lamellae, as in Hughes et al., 1979).

The objectives were almost never expressed as hypotheses regarding the mechanisms or processes governing the exposure and effects of metal mixtures. Rather, the objectives usually involved investigating whether a particular metal mixture would produce a particular effect. Occasionally, the objectives were not clearly stated. Rather, the studies appeared to be broad, exploratory exercises with a number of measures that were seldom related to their relevance to the population or community level, as identified by Farag et al. (1994).

Measures of effects varied, ranging from measures commonly used in standard toxicity tests (i.e., mortality or growth) to specific effects such as susceptibility to fungal or bacterial infections. Others used behavioral measures including prey vulnerability (Sullivan et al., 1978), schooling behavior (Sullivan et al., 1978) and locomotory activity (Gerhardt, 1990; Smith and Haines, 1995). Larger scale, population and community measures were examined by some researchers (Genter and Amyot, 1994; Horne and Dunson, 1995; Clements et al., 1989).

The application of the research to current or potential future problems was not always clear. In some papers, the introduction provided a clear rationale for the research, including the geographical location, issues of concern involving multiple stressors, and whether the issue was current or was expected to arise because of a proposed human activity (Lehtinen, et al., 1984; Farag et al., 1994; Mount et al., 1994). However, such studies were the exception.

It is critical to understand the bioavailability and actual uptake of metals by receptors in multiple-stressor studies. Metals are usually accompanied by other stressors, such as alterations in flow regime, temperature, pH, or physical habitat alterations (e.g., downstream of mine tailings dams with acid rock drainage and stream channelization). Mitigation measures should be directed towards the dominant stressors and should be done with knowledge of stressor interactions. Often, the presence of natural mitigating factors (such as high iron/manganese oxide content or high organic content) alleviates metal toxicity, as was observed in the Moira River system (Golder, 2001). If the role of these mitigating factors is not known, mitigation measures such as the dredging of metal-contaminated sediments may cause more harm than good.

Measures of effects on individuals in experimental studies can be divided into two categories: measures used in standardized single-species toxicity tests (e.g., mortality, growth, and reproductive endpoints such as the number of young/female); and measures specific to the expected toxic mechanism (e.g., histopathology, blood chemistry, susceptibility to infection, avoidance, and emergence of insects). Measures of effects on populations and communities in experimental studies are almost always limited to changes in the community assemblage (as indicated by taxa richness, presence of sensitive species, relative abundance). Measurement of functional parameters (e.g., primary or secondary productivity) is less prevalent.

The rationale for the selection of the measures used in studies of metal mixtures was rarely presented in detail. For example, the reason why the combination of arsenic, cadmium, copper, lead, and selenium studied by MacFarlane et al. (1986) would be expected to change the susceptibility of striped bass to infections by *Flexibacter* was not presented. The rationale for the selection of gross pathology, lipid peroxidation and ion regulatory dysfunction as measures of effects from a mixture of metals was that "these responses represent a variety of physiological and biochemical processes" and the authors hypothesized that some or all would be affected by the metals (Farag et al., 1994). This rationale was not based on a conceptual model about how the particular metal mixture would interact based on the metal chemistry, exposure routes, and fish physiology. Nor did the rationale present any hypotheses about whether the observed responses would represent additive, antagonistic, or synergistic effects.

No methods were observed in the literature that would conclusively determine the underlying mechanisms of metal mixture toxicity. The literature was limited to examining overall effects of metal mixtures, without "teasing apart" the effects to determine exactly how these metals are interacting to produce toxic effects.

Selected Observational Studies of the Effects of Metal Mixtures and Metals Plus Other Stressors

Although the focus of this review is on experimental studies, a few selected observational studies of the effects of metal mixtures are discussed here to illustrate some of the important findings and limitations of such studies.

The characteristics of metal-degraded systems have been the subject of considerable research over the past 20 to 25 years. The focus of this research has usually been on periphyton or benthic invertebrate communities. The most common approach is to collect information on the taxa present and their abundance and then to explore the data for statistical relationships with the physical and chemical characteristics of the water and/or sediment. This often includes a variety of metals, plus variables known to affect the bioavailability and toxicity of metals (pH, hardness of water; acid-volatile sulfides, total organic carbon, particle size of sediments).

The Sediment Quality Triad Approach (Chapman et al., 1992) uses a combination of observational data on benthic invertebrate communities, sediment chemistry, and laboratory toxicity tests to build a weight-of-evidence argument for degraded sediment communities. These studies have generated a broad understanding of the response of periphyton and benthic invertebrate communities to metals. This response can be generalized as the elimination of metal-sensitive taxa. Biomass is often not significantly affected, because metal-tolerant taxa increase in abundance in response to the reduced competition for habitat and food.

Although these observational studies have been useful in identifying the general attributes of metal-degraded systems, the actual role of particular metals versus the role of metal mixtures is difficult to deduce from these data. Furthermore, the ecological processes leading to the resulting communities are not identified. Thus, the ability to predict the response of a periphyton or benthic invertebrate community (in terms of abundance, taxa richness, or the presence of particular taxa that may be important as fish food) is limited. For example, it is usually impossible to predict with any confidence that a benthic invertebrate community will include specific keystone fish-food taxa or the required biomass to support fish in the presence of a particular list of metals. This is because the relationships identified in the literature are almost always with grouped variables. These grouped variables usually include a list of several metals, plus modifying variables such as pH or sediment texture.

The challenges of using observational studies as the basis for extrapolation to other sites with multiple metal stressors can be illustrated by a study of the effects of metal contamination in the Clark Fork River, Montana. Complementary studies using sediment and pore water toxicity tests as well as the Sediment Quality Triad addressed the impacts of past mining on the river (Kemble et al., 1994; Canfield et al., 1994). The toxicity test results showed that none of the sediments caused lethality, but that several elicited sublethal effects on growth and maturation (Kemble et al., 1994).

A complex relationship exists between sublethal effects and exceedances of individual metal "no effect concentrations." In some cases, the toxicity could not be explained by metal concentrations alone. Kemble et al. (1994) noted that the cause of sediment toxicity or the interactive effects of sediment toxicants should be determined by spiking sediment with individual chemicals or with mixtures of chemicals. Once the cause of the sediment toxicity was identified, better decisions could be made regarding remediation options.

Field data from the Clark Fork River showed that abundance of Chironomidae was significantly correlated with total concentrations of copper, cadmium, and zinc in pore water, and simultaneously extractable metal zinc in sediment. There was no way of distinguishing whether the effects of these metals were additive, dominated by one metal (such as copper), or were a complex function of the particular metal mixture and the specific physical and chemical characteristics of the Clark Fork River. The results suggested a complex relationship because the laboratory toxicity of sediment samples was associated with copper and zinc in sediment and with total ammonia in pore water. Furthermore, the toxicity data indicated that something other than metals caused toxicity in some of the sediment samples.

Finally, chironomid genera richness in the Clark Fork was higher at the more metal-contaminated sites than at reference sites. The cause of this counterintuitive result was hypothesized to be the intermittent disturbance of increased metal concentrations after high-flow events. This intermittent disturbance would keep the community at nonequilibrium. It was inferred that at equilibrium, genera richness would be lower because competitively superior species are selected for, thereby reducing the number of taxa. The hypothesized underlying ecological processes causing the higher abundance of chironomids at metal-impacted sites were increased food resources because of the lack of competitors or a reduction in predators.

The Clark Fork study was thorough, careful, and accompanied by a wealth of supporting data from related studies. It provided excellent support for a specific risk assessment; however, it did not produce results that could be extrapolated to other sites with any confidence. Instead, it generated a number of interesting hypotheses. The generation of testable hypotheses was probably the most useful aspect of observational research in multiple stressors.

4.3 Organic Chemicals (Including Pesticides); Organic Chemicals Plus Other Stressors

The literature search revealed that although there is a very large amount of literature on mesocosm studies of single pesticides, studies of pesticide combinations are sparse. The stressors, receptors and measures used in the literature on pesticide combinations are summarized in Table 4-3.

Table 4-3. Stressors, Receptors, and Measures Used in Experimental Studies of Pesticide Combinations.

Stressors	Receptors	Measures	Reference
Atrazine, chlorphyrifos, monosodium methanearsentate	*Hyla chrysoscelis* (tadpoles)	Abundance, metamorphosis, developmental, and behavioral abnormalities	Britson and Threlkeld, 1998
Alachor and atrazine	Benthic algal communities	Biovolume of viable algal cells and relative abundance of the six dominant algal species	Carder and Hoagland, 1998
Atrazine and bifenthrin	Natural plankton assemblages and bluegill	Zooplankton and phytoplankton community analyses, algal chlorophyll, primary productivity, fish mortality and basic limnological variables (temperature, DO, turbidity, total nitrogen, and phosphorus, pH)	Hoagland et al., 1993
Atrazine, alachlor, metachlor, metribuzin	Benthic algal communities	Productivity and nutrient uptake	Krieger et al., 1988
2,4-D and 2,4,5-T	*Chironomus*	Life history characteristics	Sigmon, 1979
Pentachlorophenol and hypoxia	*Haliotis rufescens* (red abalone)	Physiological (changes in intracellular pH, ATP levels, PA levels	Tjeerdema et al., 1991
Atrazine, alachlor, metolachlor, and metribuzin	*benthic invertebrates*	Productivity and nutrient uptake	Krieger et al., 1988
2,4-D and 2,4,5-T	*Chironomus*	Sublethal responses (pupation, emergence) and mortality	Sigmon, 1979

4.3.1 A Brief Synopsis of Studies of Pesticide Combinations

Carder and Hoagland (1998) assessed the combined effects of three environmentally realistic levels of two pesticides on epipelic algae. The effects of the two pesticides together on algal community biovolume were found to be additive rather than synergistic with the lack of significant interaction likely attributable to their differing modes of action. This finding has important implications for water quality criteria.

Hoagland et al. (1993) assessed the individual and combined effects of two pesticides on a natural plankton assemblage and bluegill sunfish. They observed interaction effects indicating that when either compound was introduced at ecologically realistic levels, its effects were essentially masked if the other toxicant was present at high concentrations, and that the two pesticides did not act synergistically.

Hanazato (1991) analyzed the influence of predation on the effects of pesticide application on zooplankton communities, and found that the community structure changed because of predation of zooplankton. Hossain et al. (1987) assessed the combined effects of a pesticide and a detergent on organisms from three trophic levels. The combined effects of the two pollutants were more toxic than their individual effects on all three trophic levels, suggesting a synergistic effect.

Studies of pesticides in combination with other stressors were somewhat more common than studies of combinations of pesticides. The stressors, receptors and measures used for the study of pesticides plus other stressors are summarized in Table 4-4.

Table 4-4. Stressors, Receptors, and Measures Used in Studies of Pesticides in Combination with Other Stressors.

Stressors	Receptors	Measures	Reference
Thiodan (endosulfan) and chromium	*Oreochromis mossarbilus* (fish), *Cyclops viridis* (plankton), *Branchiura sowerbyi* (worm)	Mortality, behavior, and reproduction	Sarkar and Konar, 1997
The detergent ekaline and dichlorvos	*Tilapia mossambica* (fish), *Diaptomus ferbezi* (plankton) and *Branchiura sowerbyi* (worm)	Lethality	Hossain et al., 1987
Diazinon and copper	*Ephoron virgo* (mayfly)	Mortality	Van Der Geest et al., 2000
Chlorpyrifos, chlordane, copper, cadmium	*Mysidopsis bahia* (opossum shrimp)	Mortality	Murray, 1997
Carbaryl with and without the predator *Chaoborus*	Zooplankton	Community assemblage	Hanazato, 1991
Atrazine with light and temperature	*Nannochloris oculata* and *Phaeodactylum tricornutm* (algae)	Growth	Mayasich et al., 1987
Polycyclic aromatic hydrocarbons (PAHs), polychlorinated biphenyls (PCBs), and chlorinated pesticides	*Pleoronectus vetulus* (English Sole)	Hepatic glutathione, L-cysteine, and γ-glutamylcysteine	Nishimoto et al., 1995
3 pesticides: fenitrothion 2,4-D, aminocarb in the presence of 4 different pH levels: pH 4.6, 5.6, 6.9, and 8.5	*Oncorhynchus mykiss* (rainbow trout)	Brain acetylcholinesterase and whole body aminocarb residue	Doe et al., 1988
Endrin and malathion	*Jordanella floridae* (flag fish)	Acute and chronic effects on growth, mortality	Hermanutz et al., 1985
Glyphosate and nutrients	Aquatic periphyton	Species composition and biomass accrual	Austin et al., 1991
Atrazine, chlorpyrifos, monosodium methanearsentate, and methyl mercury	*Hyla chrysoscelis* (gray treefrog tadpoles)	Abundance, growth, metamorphosis, and abnormalities	Britson and Threlkeld, 1998
Carbofuran and suspended solids	*Daphnia magna*	Impaired motility, mortality	Herbrandson et al., 1999

4.3.2 Review in the Context of Risk Assessment

The receptors in pesticide combination studies were generally algae or benthic invertebrates, with one study of frog tadpoles. Experimental studies of pesticide combination effects on fish were limited, with Hoagland et al. (1993) testing bluegill sunfish.

Standard test species were not used in studies of pesticides in combination with other stressors. There was seldom any explanation of why the particular receptors were chosen. Krieger et al. (1988) explained the use of aufwuchs in their experiments as a function of their importance as a major food source to other aquatic organisms.

The routes of exposure in studies of pesticide combinations were primarily via water (Hossain et al., 1987; Hanazato, 1991; Carder and Hoagland, 1998). One study, however, used soil slurry to simulate a sediment plume in runoff (Hoagland et al., 1993). There were no dietary exposures and no sediment exposures found in the literature. Britson and Threlkeld (1998), however, reported in their experiments of pesticides and methyl mercury on amphibians that the methyl mercury remained in the sediment for the duration of the experiment, and was not recorded in the water column. They did not expand this observation to hypothesize on methyl mercury effects via amphibians in contact with the sediments. Krieger et al. (1988) used the water exposure pathway when studying the effects of atrazine, alachlor, metolachlor, and metribuzin combinations on benthic invertebrate productivity.

The ecological assessment endpoint was rarely stated, but was mostly implied when describing the experimental results of pesticide and other stressor combination studies (Sigmon, 1979; Britson and Threlkeld, 1998; Krieger et al., 1988). Some studies examined the effect of pesticides on community structure (Hoagland et al., 1993; Carder and Hoagland, 1998). The objectives of the studies were very specific to the expected effects on the receptors because of the type of pesticide (e.g., herbicides were expected to reduce primary productivity). The objectives were not expressed as hypotheses regarding the mechanisms or processes governing the exposure and effects of pesticide mixtures. Rather, the studies appeared to be primarily exploratory (Sigmon, 1979; Britson and Threlkeld, 1998; Krieger et al., 1988). Exceptions to this trend included Britson and Threlkeld's (1998) study of abnormalities in amphibians caused by exposure of agrichemicals and methyl mercury. Their objectives were not expressed as hypotheses regarding the mechanisms or processes governing the exposure and effects of pesticide mixtures. Rather, the objectives involved investigating whether a particular pesticide mixture would produce a particular effect on a receptor for which the pesticide was not intended.

Measures of effects were highly variable and appeared to be related to a combination of the type of receptor chosen for the experiment and the type of pesticides tested. For example, if herbicides were tested on algae, then biomass or productivity were the endpoints. In some cases, the link between the pesticide and the endpoints was not obvious and was not explained (Sigmon, 1979).

The receptors in studies of pesticides in combination with other stressors included a wide variety of fish species, periphyton, phytoplankton, zooplankton, benthic invertebrates and gray treefrog tadpoles. This wide assortment of receptors reflects the rather eclectic mix of studies, the wide range of receiving environments of concern and the various research interests of the authors.

Measures of effects were highly variable and appeared to be related to a combination of the type of receptor chosen for the experiment and the type of stressors tested. For example, Nishimoto et al. (1995) hypothesized that the combination of PAHs, PCBs and chlorinated pesticides was related to effects on a number of liver proteins.

4.3.3 A Brief Synopsis of Studies of Other Organic Compounds and Organic Compounds Plus Other Stressors

Experimental studies of organic compounds other than pesticides usually involved chlorinated organic compounds (notably PCB and pentachlorophenol). Other classes of organic compounds (such as the PAHs) are rarely, if ever, studied in combination, except in uncharacterized mixtures. The limited literature on mixtures of organic compounds and on combinations of organic compounds plus other stressors is summarized in Table 4-5.

Table 4-5. Stressors, Receptors, and Measures Used in Studies of Organic Compounds in Combination with Other Stressors.

Stressors	Receptors	Measures	Reference
PCB and arsenic	*Chlamydomonas* (green alga)	Inhibition of growth	Christensen and Zielski, 1980
β-Napthoflavone and resin acids	*Salmo salar* parr (Atlantic salmon)	Biotransformation enzymes	Croce et al., 1995
Pentachlorophenol and hypoxia	*Haliotis refescens* (abalone)	Physiological (inorganic monophosphate, phosphoarginine (PA), and adenosine triphosphate (ATP) concentrations in muscle, intracellular pH change	Tjeerdema et al., 1991
Pentachlorophenol and temperature	*Haliotis refescens* (abalone)	Physiological (inorganic monophosphate, PA, and ATP concentrations in muscle, intracellular pH change	Tjeerdema et al., 1993
3,4-Dichloroaniline with different nutrient levels	*Daphnia magna* and other naturally occurring freshwater zooplankton species	Population densities of zooplankton species and phytoplankton biomass as determined by chlorophyll a concentrations	Jak et al., 1998
Nonylphenol, bisphenol a	*Xiphophorus helleri* (swordtail fish)	Endocrine disruption	Kwak et al., 2001
Thiolate organic compounds and methyl mercury	Sheepshead minnow	Accumulation and redistribution of methyl mercury	Leaner and Mason, 2001
Salinity, temperature and oil	*Fundulus heteroclitus* (killifish)	Survival and developmental rate of embryos	Linden et al., 1979
Oil and oil dispersant mixtures	Zooplankton and benthos of freshwater ponds	Relative abundance and species presence/absence	Scott et al., 1984
Pulp mill effluents and nutrients	invertebrate communities	Community structure	Podemski and Culp, 1996
Anthracene and sunlight	*Lepomis macrochirus* (bluegill sunfish)	Acute mortality	Bowling et al., 1983
Pulp mill effluents	Periphyton	Community structure	Amblard et al., 1990
Pulp and paper mill effluents and hydroelectric dams	*Catostomus commersoni* (white sucker)	Fish health	Munkittrick et al., 2000

Stressors	Receptors	Measures	Reference
Cadmium, PCB (aroclor 1254)	*Parophrys vetulus* (English Sole)	Serum constituents, histological impact	Rhodes et al., 1985
Dissolved organic carbon (DOC), cadmium, copper	*Oncorhynchus mykiss* (rainbow trout)	Respiratory and ionic regulation	Richards et al., 1999
Pulp mill and sewage effluent and low temperature	*Baetis tricaudatus* (mayfly)	Mortality	Lowell and Culp, 1999
Thiodan and chromium	*Oreochromis mossambicus)*(fish), *Branchiura sowerbyi* (worm) and *Cyclops viridis* (plankton)	Mortality	Sarkar and Konar, 1997
Phenol and ammonia	*Baetis rhodani* (mayfly)	Survival	Khatami et al., 1998
Triphenyl phosphate and sediments	Periphyton and invertebrate communities	Abundance, drift, nutrient dynamics, leaf decomposition	Fairchild et al., 1987
PCB with nutrient enrichment	Phytoplankton	Growth	Lin and Simmons, 1981
Pulp mill effluent and food levels	*Baetis tricaudatus* (mayflies)	Survival, growth, molting and morphological development	Lowell et al., 1995a
Pulp mill and municipal effluents, decreased oxygen levels	Benthic biota	Growth, benthic community structure	Lowell et al., 2000
Pulp mill, municipal and oil sands effluents, agricultural runoff	*Lota lota* (burbot), *Esox lucius* (northern pike), *Platygobio gracilis* (flathead chub)	Sex steroids, mixed-function oxidase, hepatic vitamins, metallothionein, sexual maturity, external abnormalities, stable isotopes	Cash et al., 2000

Jak et al. (1998) examined the sensitivity of one zooplankton species to a toxicant compared to other zooplankton species at different nutrient levels. Both the effects of the toxicant on zooplankton and the diminished control exerted by zooplankton on phytoplankton were intensified under nutrient rich conditions.

Lin and Simmons (1981) examined the combined effects of PCB and nutrient enrichment on a natural phytoplankton community. The phytoplankton production stimulated by nutrient enrichment was reduced by the presence of PCB, which resulted in differential effects on species growth rates.

Culp et al. (2000) examined the interactive effects of nutrients and contaminants from pulp mill effluents on riverine benthos, a study that could not be accomplished using only field biomonitoring data. Bleached Kraft mill effluent (BKME) concentrations equivalent to naturally occurring levels did not cause measurable toxicity, but rather, BKME associated increases in phosphorus produced enrichment effects including increases in periphyton and insect biomass, and invertebrate abundance.

Haley et al. (1995) and Hall et al. (1991) examined the effects of BKME on several trophic levels (rainbow trout, periphyton, and macroinvertebrates). The studies measured the effluent quality (BOD, color, total suspended solids, and turbidity) resulting from mill-process changes and correlated these data with measurements of periphyton primary production, benthic macroinvertebrate biomass, and population diversity and rainbow trout growth, survival, and production.

Lowell et al. (1995b) conducted an experiment to determine the effects of current velocity on sensitivity of a mayfly to toxicants (sodium chloride). Their results indicated that differences in current velocity did not have a great effect on the sensitivity of the mayfly to the toxicant, despite additional stress induced by low-current velocity (thought to be a result of insufficient delivery of oxygen to the mayfly's gills).

4.3.4 Review in the Context of Risk Assessment

The receptors in studies of organic compounds in combination with other stressors include a wide variety of fish species, periphyton, phytoplankton, zooplankton, and benthic invertebrates. As with pesticides, this wide assortment of receptors reflects the range of compounds, the variety of receiving environments of concern, and the various research interests of the authors.

Standard test species were rarely used in studies of organic compounds in combination with other stressors. There was seldom any explanation of why the particular receptors were chosen. In some cases, receptors were taken from the study areas to the laboratory or field mesocosm. The choice of receptors was therefore implied to be a representation of the receptors found in the study lake, river, or stream (Christensen and Zielski, 1980; Podemski and Culp, 1996; Amblard, et al., 1990; Croce et al., 1995). In one case, the receptor was chosen to demonstrate its effectiveness in determining endocrine-disrupting compound effects (Kwak et al., 2001). Tjeerdema et al. (1991, 1993) described the link between the test species, *Haliotis rufescens,* and the stressors being tested. They stated that the concentrations and combinations of stressors were chosen to replicate those found in the intertidal zone environment to determine stressor effects (Tjeerdema et al., 1991, 1993).

The exposure of receptors to the organic compounds was expressed as concentrations in water. A number of studies chose field relevant levels of stressors (Newman and Perry, 1989; Lowell et al., 1995a; Culp et al., 2000).

The ecological assessment endpoint was rarely stated. There are a number of studies that looked at the effects on community structure (Jak et al., 1998; Scott et al., 1984; Amblard et al., 1990; Podemski and Culp, 1996; Lowell et al., 2000). As with pesticide studies, the objectives were not expressed as hypotheses regarding the mechanisms or processes governing the exposure and effects of pesticide mixtures. Rather, the studies appeared to be primarily exploratory.

Measures of effects were highly variable. In some cases, the endpoints were related to known effects of the compounds in other species (i.e., nonylphenol and bisphenol A have been shown to be endocrine disrupters in a number of studies; Kwak et al., 2001). Croce et al. (1995) used physiological changes such as hyperbilirubinaemia when examining effects of ß-naphthoflavone and resin acids on Atlantic salmon. Tjeerdema et al. (1991, 1993) used the endpoints of altered PA, ATP, inorganic phosphate levels, and intracellular pH when examining the effects of pentachlorophenol plus either hypoxia or temperature on abalone. The choice of respiratory and ionic regulation when two metals plus the mediating factor for metal interactions with gill (DOC) were studied is related to the known effect of metals on gill membrane structure and ion exchange (Richards et al., 1999). In other cases, general nonspecific endpoints such as mortality, growth, and species richness/diversity were used (Bowling et al., 1983; Amblard et al., 1990; Podemski and Culp, 1996).

Some studies focused on the effects expected because of the particular effects of the pesticides. For example, a combination of the insecticide esfenvalerate and the herbicide atrazine was studied for direct mortality effects on fish and zooplankton (because of known high toxicity of esfenvalerate), direct effects on photosynthesis, community respiration, and net primary production (caused by atrazine), and indirect effects because of the change in habitat structure caused by effects on the plant community (La Point and Fairchild, 1994). In an interesting and sophisticated study of a starch blended with vinyl alcohol copolymers being developed as a substitute for nonbiodegradable plastics, Klein-MacPhee et al. (1993) measured mortality of larval flounder, microzooplankton abundance, larval growth, larval gut contents, and DO concentrations. These measures addressed the possibility of direct toxicity of the starch blend or its degradation products, ingestion of particles of the degrading starch blend by larval flounder, and BOD created by the decomposition of the starch blend. This study was one of the few that explicitly related hypothesized effects to particular measures.

4.4 Nutrients, Chlorine, Temperature, and Combinations with Other Stressors

4.4.1 A Brief Synopsis of the Studies

A number of studies focused on the effects of nutrients in combination with other stressors such as pulp mill effluents, metals, suspended solids, and temperature (Table 4-6). The most common receptors were periphyton and benthic invertebrate communities. The most common measures were various community metrics such as taxa richness, abundance, and the presence of particular taxa.

Table 4-6. Stressors, Receptors, and Measures Used in Studies of Nutrients or Major Ions in Combination with Other Stressors.

Stressors	Receptors	Measures	Reference
Suspended clay and phosphorus	Phytoplankton	Abundance, primary productivity	Cuker, 1987
Nutrients and pulp mill contaminants	Periphyton and benthic invertebrates	Periphyton biomass, benthic insect density and composition, biomass of abundant invertebrates	Podemski and Culp, 1996
Nutrients and pulp mill contaminants	Benthic invertebrates and periphyton	Periphyton biomass, benthic insect density and composition, biomass of abundant invertebrates	Culp and Podemski, 1996
Nutrients and pulp mill contaminants	Benthic invertebrates and periphyton	Periphyton biomass, benthic insect density and composition, biomass of abundant invertebrates	Culp et al., 2000
Nitrite, pH and chloride	*Lepomis macrochirus* (bluegill)	Mortality	Huey et al., 1982
Nutrients and trace elements (mixture of Cu, As, Cd, Ni, and Zn)	Estuarine food web	Production, abundance, growth, community assemblage	Breitburg et al., 1999
Arsenic, cadmium, copper plus nitrogen, and phosphorus	*Macoma balthica* (clams), *Crassotrea virginica* (oysters) and *Fundulus heteroclitus* (mummichogs)	Trace metal uptake in bivalves and fish	Connell et al., 1998
Nutrients, trace metals	Periphyton	Photosynthesis and respiration	Laursen et al., 1999
Metals (copper, arsenic and cadmium) and nutrients (nitrogen, phosphorus)	Estuarine ecosystem	Community assemblage, growth, production	Osman et al., 1999
Phosphate and manganese	Phytoplankton	Growth	Shapiro and Glass, 1975
Nutrients and temperature	Periphyton	Community assemblage, biomass production	Brown, 1981
"Agricultural pollution"	Stream community	Community assemblage	Kownacki et al., 1985
Nutrients and sedimentation	Benthic invertebrates	Community assemblage	Lemly, 1982
Phosphorus loading reductions plus zebra mussels	Phytoplankton	Community assemblage	Nicholls and Hopkins, 1993
pH, chloride, and nitrite	*Lepmis macrochirus* (bluegill)	Mortality	Huey et al., 1982
Chlorine and temperature	*Salvelinus fontinalis* (brook trout)	Mortality	Thatcher et al., 1976

Stressors	Receptors	Measures	Reference
Arsenic, copper, cadmium plus phosphorus and nitrogen	Phytoplankton communities	Community assemblage	Sanders et al., 1999
Nutrients and Cd, Se, Zn	Marine phytoplankton (four species)	Metal assimilation	Wang and Dei, 2001
Sulfate and desiccation	Ecosystem	Nutrient dynamics	Lamers et al., 1998
Agriculture, urbanization and climate	Phytoplankton and invertebrates	Community dynamics	Hall et al., 1999
Nutrients and herbivory (to simulate exotic species)	phytoplankton	species richness and diversity	Proulx et al., 1996
Sodium chloride and current velocity	*Baetis tricaudatus* (mayflies)	immobilization and number of molts	Lowell et al., 1995b
Temperature and chlorine	Estuarine fish	Survival of young-of-the-year	Hoss et al., 1977
Chlorine and ammonia	*Potamogeton crispus* L.	Litter breakdown	Newman and Perry, 1989
Chlorine and ammonia	*Potamogeton crispus* L., microbes and macrobenthic "shredders"	Litter decomposition, changes in microbe and shredder densities	Perry and Troelstrup, 1988
Chlorine and temperature	*Salvelinus fontinalis*	Mortality	Thatcher et al., 1976

Chlorine has been frequently studied in combination with other stressors (Table 4-6). It can be assumed that the effects of chlorine plus other stressors is of interest because many complex effluents contain residual levels of chlorine. The interaction between chlorine and temperature has been the subject of some experimental work (Table 4-6). These studies would be particularly relevant to effluent discharges that are both warmer than ambient temperatures and contain residual chlorine (e.g., municipal effluents, thermal power plant discharges where chlorine treatment for zebra mussels is required). Chlorine is acutely toxic at very low concentrations; therefore, mortality is often one of the measures used in experimental studies of chlorine plus other stressors (Table 4-6).

Newman and Perry (1989) investigated the combined effects of chlorine and ammonia on litter breakdown. They found that litter decomposition was significantly reduced in treatments containing both chlorine and ammonia in comparison with treatments of just chlorine, suggesting that chlorine plus ammonia combinations were more toxic than chlorine alone.

There is a notable lack of experimental studies of combinations of stressors associated with agriculture, except for uncharacterized mixtures such as "agricultural pollution." Only one experimental study of nutrients plus sedimentation was found (Lemly, 1982). No experimental studies of nutrients plus sedimentation plus pesticides were found, despite the common occurrence of this stressor combination in aquatic ecosystems.

A common finding among these papers is the lack of thorough exploration of the nature of the interaction between different stressors. Several studies indicated that nutrient enrichment increased or decreased metal uptake or toxicity, without any exploration of possible mechanisms or even a clear attempt to ascribe sources (Murray, 1997; Connell et al., 1998; Osman et al., 1999; Laursen et al., 1999). Sanders et al. (1999) commented on an interaction between phosphorus concentration and arsenic assimilation by phytoplankton, which may be a simple competition effect. In many of these studies, the added nutrients may be considered a stimulant of growth and vigor (and potentially resistance to toxins) rather than a stressor in the strict sense. They would be considered subsidies in the scheme of Odum (1985). Consequently, detailed examinations aimed at untangling the effects and interactions of two chemical stressors were almost never completed.

4.4.2 Review in the Context of Risk Assessment

The receptors in studies of nutrients, chlorine, temperature, and combinations of other stressors included a variety of fish species, periphyton, phytoplankton, zooplankton, and benthic invertebrates. As with pesticides, this assortment of receptors reflects the range of compounds, the variety of receiving environments of concern, and the various research interests of the authors.

Standard test species were rarely used. There was seldom any explanation of why the particular receptors were chosen. In some studies, the research concerned a particular site (i.e., river, lake, and estuary) that was known to be suffering pollution from metals and nutrients (Connell et al., 1998; Breitburg et al., 1999; Osman et al., 1999).

Some of these studies provided explanations for the selection of the receptors. Cuker (1987) studied the effect of phosphorus and suspended clay additions on primary productivity and phytoplankton community structure because both factors have been shown to have differing effects on these response variables in the past. Net community productivity was stimulated in treatments containing phosphorus, regardless of the presence of clay. In addition, phosphorus additions changed algal community structure to domination by cyanobacteria whereas treatments with clay contained mostly flagellates. Lamers et al. (1998) studied the response of nutrient dynamics, redox potential, and pH in wetland pore soil to changes in sulfate concentration and desiccation. Both of these factors affected soil pore water acidity, whereas acidity and the presence of oxygen altered redox potentials and substantially affected nutrient dynamics.

Benthos (periphyton and macroinvertebrates) are the most common receptors in these studies because they account for much of the energy flow through these systems (Culp et al., 2000; Lowell et al., 2000). The functional diversity within the benthos allows both allochthonous and autochthonous inputs to be processed efficiently and transformed into biomass available to higher trophic levels (Lemly, 1982). As such, changes in these lower trophic levels will quickly be reflected in higher ones.

The routes of exposure were via water (Cuker, 1987; Lamers et al., 1998; Huey et al., 1982; Breitburg et al., 1999; Connell et al., 1998; Laursen et al., 1999; Osman et al., 1999). The measures of exposure were expressed in concentrations (Lamers et al., 1998) and also loading rates (mg P m^{-2} d^{-1}; Cuker, 1987).

Exposure to nutrients is generally measured as the concentrations of the individual nutrients, but nutrient ratios may prove to be useful in assessing or predicting the interaction of multiple stressors. For example, Breitburg et al. (1999) performed mesocosm experiments that suggested that the N:P and its implications for nutrient limitation can affect the interaction of trace metals and nutrients. When N:P was low, and therefore nitrogen limited phytoplankton production, they hypothesized that the excess phosphorus reduced the toxic effects of arsenic (Breitburg et al., 1999).

The assessment endpoints for these studies are more often implicit than explicitly described. Studies of nutrients plus other stressors are important because effluent discharges commonly combine elevated nutrient concentrations with a variety of potentially toxic constituents (e.g., pulp mill effluents, municipal discharges). In some cases, the authors make it clear that they are interested in the overall effect of stimulatory constituents such as nutrients and toxic constituents because of the relevance to regulation of discharges (Culp et al., 2000; Hall et al., 1991; Haley et al., 1995). However, in other cases, the authors do not clearly describe the assessment endpoints or the context for the research.

Measures of effects ranged from standard endpoints such as mortality to measures of community structure or function. There was one case of a population-level endpoint (survival of young-of-the-year fish) (Hoss et al., 1977).

Many studies of combinations of nutrients with other stressors use measures of direct effects such as mortality and measures of community structure. The focus on community structure is common because of the large literature and reasonably good understanding of the changes to expect in the relative abundance of various taxa with nutrient enrichment.

Despite the fact that nutrients directly affect productivity, primary and secondary productivity are seldom measured, with some notable exceptions (Cuker, 1987, Lamers et al., 1998). The reasons for this are not usually given; however, there appears to be an implicit assumption that community structure is a more sensitive indicator of nutrient-related effects than productivity because of built-in functional redundancy in most ecosystems. This assumption would be in accordance with Odum's model of ecosystem stress (1985), which suggests that structural elements of ecosystems are more sensitive indicators of stress than are process-oriented functional attributes because species replacement and other adjustments keep the overall function of the ecosystem steady until the stress is overwhelming. This assertion is supported by the work of Schindler (1987) and Breitburg et al. (1999, see below).

The emphasis on community structure is seen in studies of combinations of nutrients and other stressors such as temperature, salinity, sedimentation, and pulp mill effluents (Culp and Podemski, 1996; Culp, Cash, et al., 2000a; Huey et al., 1982; Osman et al., 1999; Shapiro and Glass, 1975; Brown, 1981; Kownacki et al., 1985; Lemly, 1982; Nicholls and Hopkins, 1993).

In nutrient studies, measures of ecosystem function such as net community productivity (Cuker, 1987) and nutrient dynamics (Lamers et al., 1998) have been used. Several physical parameters also were measured in these studies, such as conductivity, pH, temperature, oxygen, and redox potential. One study dealing with the effects of climate change, urbanization, and agriculture on lake eutrophication measured lake sediment fossil data on phytoplankton and chironomid community dynamics (Hall et al., 1999). This primarily was done because data on the historical changes in ecosystem variables needed to address their objective were impossible to obtain.

Studies of nutrients plus trace metals have used functional measures of productivity and metal uptake (Breitburg et al., 1999; Laursen et al., 1999; Connell et al., 1998). Breitburg et al. (1999) examined functional responses as well as changes in community structure in response to additions of nutrients and a mixture of trace metals to estuarine mesocosms. Temporal variability greatly affected stressor interactions because of variation in species abundances and responses to stressors as well as the levels of the individual stressors. Algal taxonomic structure was affected by trace metal addition, but species complementarities buffered the reduction in productivity. Metal additions had less impact when seasonal differences in nutrient availability resulted in lower N:P ratios. In general, this study supported the assertion that functional responses are less sensitive to perturbation than structural ones.

4.5 Habitat Change, Invasion of Nonindigenous Species, Flow Regulation, Overharvesting, and Global Climate Change

This grouping of stressors is adopted from the list of the top-priority stressors identified by the Integrated Risk Project of the U.S. EPA (Harwell and Gentile, 2000). The project proposed that the highest ecological risks to the U.S. derive from habitat alteration, hydrologic modification, over-harvesting, invasive exotic species and global climate change (Harwell and Gentile, 2000). Given this ranking, it is interesting to find so few publications on the effects of these stressors in combination with each other, or with stressors such as metals and pesticides (Table 4-7). Of course, it is difficult to study stressors such as these in an experimental setting. This probably explains the relative scarcity of studies.

Table 4-7. Stressors, Receptors, and Measures Used in Studies of Habitat Change and Invasion of Nonindigenous Species.

Stressors	Receptors	Measures	Reference
Nutrient enrichment with vegetated and degraded stream banks	River fish community	Community assemblage	Growns et al., 1998
Habitat degradation	Fish community	Community assemblage	Karr et al., 1985 (literature review)
Competition, predation and disturbance (drying out of tanks)	*Rana utricularia, Scaphiopus holbrooki, Bufo americanus,* and *Hyla chrysoscelis* (amphibians)	Community structure	Wilbur, 1987
Exotic species (literature review)			Mills et al., 1994
Low pH and a leech predator	*Macrobdella decora* (leech), *Rana sylvatica, R. clamitans, R. catesbeiana, R. pipiens* (frogs)	Hatching success, hatch timing and hatch length	Schalk, 1997
Nutrients and herbivory (to simulate exotic species)	Phytoplankton	Species richness and diversity	Proulx et al., 1996

The main source of information on the effects of combinations of habitat change, invasive species, hydrologic modifications, and overharvesting is from the search for pattern in large monitoring databases. These studies will continue to be important in developing our understanding of these stressors.

In one example of an exploratory study, Hall et al. (1999) used a multivariate approach to address temporal changes in lake eutrophication by collecting historical data on 83 explanatory variables falling into three major categories: climate, resource use, and urbanization. To obtain historical data on the trophic status of the lake, sediment cores were aged using lead isotopes and analyzed for fossil phytoplankton and chironomid communities.

There have been few studies so far on the effects of large-scale climate change on aquatic ecosystems, and even fewer that look at the combined effect of climate change and other stressors. Although considerable work has been done on the effects of warmer temperatures and elevated CO_2 on production and decomposition in forests, this work has not been extended to aquatic ecosystems. Most aquatic studies of climate change consider extreme events, especially hurricanes (McDowell et al., 2000), which are anticipated to become more common as the earth warms.

Climate change encompasses a wide range of atmospheric conditions. This potential stressor can be made manageable by considering the specific effects it will have on aquatic ecosystems. Over most of North America, global warming is predicted to produce an increase in mean annual temperature, a reduction in precipitation, and an increase in the frequency and severity of extreme weather events, such as the hurricanes discussed above. Hence, we can examine the effects of climate change by studying effects of elevated temperatures and reduced water flow on aquatic ecosystems. Physical models can be used to predict the combined effect of a given reduction in rainfall and increase in mean temperature (including subtleties like the proportions of rain and snow) on runoff into a stream or lake. Similar models can predict the temperature change induced by a reduction in base flow and elevation of air temperature. It would then be possible to design experiments mimicking these changes in mesocosms or artificial streams, in which the effects on the biota would be measured. To our knowledge this has not yet been done for any climate change scenarios, let alone in combination with other stressors.

Climate change may have indirect effects on aquatic ecosystems, such as a change in primary production (both within waterbodies and in riparian vegetation so important for streams) in response to elevated temperatures and CO_2 concentrations. These changes may be minor compared with direct effects on runoff and water temperature. More problematic to study are biological consequences such as range extensions and invasions by species previously restricted by low temperatures, and possible changes in food quality and energy flow following from a warmer climate. Research into combined effects of such possibilities has only recently begun.

Proulx et al. (1996) examined the combined effects of nutrient inputs and the alteration of a lentic food web on phytoplankton species diversity. The food web treatment was the introduction of a zooplanktivorous fish (*Phoxinus eos*), which simulated decreased herbivory because of the introduction of an exotic secondary consumer. This case demonstrates the difficulties that may be encountered when measuring exposure levels involving cascading trophic interactions. Although the exotic fish was the cause of the stress, it was the cascading effect (increased herbivory) that actually acted on the receptor. Responses may also be complicated in that the exposure to herbivory depends not only on the top-down control by the exotic predator but is also subject to bottom-up control as nutrients are added and phytoplankton biomass increases.

The measures of exposure in the Growns et al. (1998) study of the combined effects of habitat degradation and nutrient enrichment were nutrient concentrations and loads in the river and bank types. Thus, the "exposure" to habitat change was represented by a discrete, descriptive variable that divided the study area into degraded and nondegraded habitat.

Proulx et al. (1996) demonstrated that the effect of exotic species on native communities and ecosystems is likely to be dependant on interactions with other ecosystem stressors. In lake enclosures, they found that although increased nutrient loads consistently reduced algal diversity, the impact of herbivory was dependant on nutrient levels. Herbivory reduced diversity under low-nutrient conditions, but increased diversity at high-nutrient loads.

CHAPTER 5.0

CONTRIBUTION OF THE CURRENT LITERATURE TO THE CONSTRUCTION OF GENERALIZED CONCEPTUAL MODELS OF MULTIPLE-STRESSOR EFFECTS

5.1 Summary Review in the Context of Risk Assessment

The current literature provides, at best, a series of site-specific glimpses of the response of ecological communities and ecosystems to multiple stressors. The definition of the problem provided by authors is usually very focused on a particular site or a particular combination of stressors. There is seldom, if ever, any reference to a more generalized model of multiple-stressor effects apart from the discussion of the expectations regarding additivity versus synergism. Many articles that purport to be multiple-stressor studies do not go beyond an inventory of the various stresses and upsets affecting the ecosystem, without even attempting to assess the interactions among them (Bailey et al., 1996; Boët et al., 1999). Even the regional risk-ranking model of risk assessment championed by Landis and colleagues (Landis et al., 1998, 2000; see Section 4.6) merely adds up all the stresses affecting different areas within the region, without attempting to compute their interactive effects.

The exceptions to these studies are papers that define the problem in terms of building a weight-of-evidence case for cause/effect relationships among stressors and receptors (Culp Lowell et al., 2000a,b), or papers by Munkittrick and colleagues that refer to a suggested general response pattern of fish to stressors (Munkittrick and Dixon, 1989; Gibbons and Munkittrick, 1994). However, even these papers seldom place the definition of the problem in a wider framework provided by a general model such as Odum's (1985) model of ecosystem stress.

Lewis (1999) provides an exception, in that his brief report of work in progress posits three testable hypotheses concerning general responses of ecosystems to stress. They are: 1) a given biotic compartment in an incrementally stressed ecosystem will show suppression of biodiversity first and suppression of biomass subsequently (similar to the idea expressed by Schindler, 1987); 2) suppression of ecosystem function is primarily caused by reduced biomass, rather than loss of diversity itself; and 3) suppression of biomass differs among biotic compartments according to a predictable series of thresholds (i.e., multicellular organisms are more sensitive than unicellular autotrophic algae that are more sensitive than microbial decomposers).

The third hypothesis implies a predictable sequence of ecosystem processes that will be lost as the severity of stress increases. Lewis's work is seminal in that it goes beyond simply listing the responses of a particular system to a particular set of stressors, but instead provides a general, testable model of ecosystem behavior. Unfortunately, work to test these hypotheses has only just begun, although Lewis (1999) suggests that preliminary studies on streams suffering acid mine drainage are consistent with his hypotheses.

Although the importance of building a relationship between stressors and receptors is of great importance, so too is the correct choice of receptors to study. The usefulness (or lack thereof) of the selected receptors as suitable indicators of the community or ecosystem is seldom critically examined. There appears to be at least two or three "camps" with respect to the most suitable and relevant receptors:

- The periphyton and benthic invertebrate camp—these researchers cite the importance of the algal or invertebrate communities to the overall structure and function of the ecosystem and to the fact that they are primarily sedentary (thus making exposure assessment easier) (Krieger et al., 1988; Culp et al., 2000; Lowell et al., 2000);
- The fish camp—these researchers cite the socio/economic importance of fish, and state that fish populations integrate site-specific factors and act as an indicator of the overall state of the ecosystem (Munkittrick and Dixon, 1989; Gibbons and Munkittrick, 1994); and
- The "maximum accumulator" camp—these researchers emphasize the importance of studying the organisms most likely to bioaccumulate the stressors of concern; these researchers are usually greatly influenced by their experiences with persistent chemicals that, at the very least bioaccumulate and often biomagnify in food chains (Ekelund, 1989; Connell, 1990; Gobas et al., 1999).

The above camps refer to stream studies. Researchers of lentic systems may also emphasize the importance of including the plankton community (phytoplankton and zooplankton) because of their important role in lake and pond food chains.

The current literature primarily relies on the traditional measures of exposure in aquatic systems (i.e., concentrations in water and/or sediments). These measures are sometimes supplemented with measures of tissue residues. Tissue residues provide a direct measure of uptake and "dose" to the organism and avoid the problems of having to predict the bioavailability of the stressors. However, because most of our knowledge of the effects of stressors on aquatic organisms is based on concentration/response relationships and not dose/response relationships, it is still difficult to extrapolate from the few studies that have measured tissue residues to other sites, or to relate the results to water or sediment quality criteria.

Exposure to nonchemical stressors such as habitat degradation and flow regulation is generally qualitative. The change in habitat or flow regime is described relative to conditions before disturbance. There are some standardized measures of habitat quality available for fish species, such as the Habitat Evaluation Procedure models developed by the U.S. Fish and Wildlife Service (U.S. FWS). Therefore, the change in habitat quality for fish can be compared from one study to another. There are also some standardized approaches to the understanding of changes in flow regime relative to fish requirements (the instream flow needs methods developed for fish by the U.S. FWS). These tools were not referred to in the multiple-stressor literature, presumably because they are not geared to unravelling the various sources of habitat degradation.

Instream flow models and fish habitat indexes only quantify the quality of the habitat, they tell us nothing about how the habitat came to be that way.

When multiple-stressor studies combine chemical and nonchemical stressors, the cumulative exposure to these stressors is difficult to describe. The current literature simply lists the chemical exposure (as concentration or tissue residue) and the relative changes in nonchemical stressors (such as relative change in habitat). A true cumulative exposure measure may involve a qualitative ranking of exposure that is derived from a sum of "exposure scores" for each stressor. Each exposure score would be derived from an evaluation of the severity, duration, and spatial extent of exposure to each stressor. For example, exposure to mercury plus habitat degradation could be assigned a rank depending on the body burden of mercury and the change in the Habitat Suitability Index (Schamberger et al., 1982) for the receptor species.

Landis et al. (1998; 2000) have proposed a scheme broadly similar to this idea. The Rank Risk Model for ERA has been devised as a method of quantifying risk in regional settings where there are multiple stressors, receptors, and possible impacts. The method consists of identifying the stress sources and habitats in different locations within a region, and then ranking their importance at each location. The number of possible risk and habitat combinations depend on the number of each identified, and some of them may overlap. The Rank Risk Model incorporates a system of ranks and weighting factors that help quantify the risk to each habitat from a number of different stressors. These weighting factors are unitless, so that risk measured in entirely different ways can still be combined. The net result is a list of risks to the various habitats in rank order by priority and weighted according to the severity of the stressor, the sensitivity of the receptor and the number of habitats affected.

The rationale for the selection of exposure pathways in multiple-stressor research is rarely put in a broader context that would contribute to a generalized model of "critical pathways" that may exist when more than one stressor is present. There is a need for a wide-ranging exercise that produces hypothesized critical pathways in the presence of the priority stressors. This exercise may reveal unique critical pathways that would not have been thought about before. For example, there is voluminous literature on the effects of various modifying factors such as water hardness or DOC on the bioavailability and toxicity of metals. This knowledge, however, has not been combined with our knowledge of the effects of increased salinity to build a conceptual model of the critical pathways of exposure in a lake that receives saline mine tailings effluent along with elevated metals. Would the salinity eliminate certain algal species that are critical to the maintenance of DOC concentrations, which in turn reduce metal toxicity? What other scenarios may produce greater exposures than would be expected if only one stressor was present? Which scenarios would be the most likely to produce increased exposure? These questions would generate testable hypotheses and would provide the rationale for the particular exposure pathways examined in the research.

Measures of effects on individuals in experimental studies can be divided into two categories. The first category is measures used in standardized single-species toxicity tests (e.g., mortality, growth, and reproductive endpoints such as the number of young/female). The second category is measures specific to the expected toxic mechanism (e.g., histopathology, blood chemistry, susceptibility to infection, avoidance, and emergence of insects).

Measures of effects on populations and communities in experimental studies are almost always limited to changes in the community assemblage (as indicated by taxa richness, presence of sensitive species, relative abundance). Measurement of functional parameters (e.g., primary or secondary productivity) is more rare.

No methods are observed in the literature that would conclusively determine the underlying mechanisms of mixture toxicity. The literature is limited to examining overall effects of mixtures, without teasing apart the effects to determine exactly how stressors are interacting to produce toxic effects. An exception to this observation is the interesting and sophisticated study of a starch blended with vinyl alcohol copolymers being developed as a substitute for nonbiodegradable plastics by Klein-MacPhee et al. (1993). These authors used measures of direct toxicity but also included measures of indirect toxicity caused by degradation products or by BOD. This study was one of the few that explicitly related hypothesized effects to particular measures.

Authors seldom discuss the relative merits of various measures with respect to their specificity or sensitivity to a particular combination of stressors. If the problem formulation for multiple-stressor research were placed within a broad model of the response of ecosystems to stress, it is possible that seldom-used or completely new measures may be revealed as being the most appropriate for multiple-stressor work. For example, the size distribution of the primary producers may shift to smaller taxa, with a subsequent cascading effect throughout the food web (Odum, 1985; Guanguo, 1990).

The above deficiencies are even more pronounced when nonchemical stressors are involved. This is because the approaches, exposure measures and effects measures commonly used in toxicology may not be applicable to nonchemical stressors such as habitat degradation or flow regulation. The interaction between chemical and nonchemical stressors may involve unique pathways of exposure (direct and indirect) and may require the use of new or less frequently used measures.

The assessment endpoints for the experimental studies of multiple stressors are more often implicit than explicitly described. In some cases, the authors make it clear that they are interested in the overall effect of stimulatory constituents such as nutrients and toxic constituents because of the relevance to regulation of discharges. However, in other cases, the authors do not clearly describe the assessment endpoints or the context for the research. The objectives of most studies are usually particular research questions (e.g., whether essential metals and nonessential metals produced different effects on gill lamellae, as in Hughes et al., 1979).

The objectives are almost never expressed as hypotheses regarding the mechanisms or processes governing the exposure and effects of multiple stressors. Rather, the objectives usually involve investigating whether a particular mixture would produce a particular effect. Occasionally, the objectives are not clearly stated. Rather, the studies appeared to be broad, exploratory exercises with a number of measures that are seldom related to their relevance to the population or community level.

The application of the research to current or potential future problems is not always clear. In some papers, the introduction provides a clear rationale for the research, including the geographical location, issues of concern involving multiple stressors and whether the issue is current or expected to arise because of a proposed human activity. However, such studies are the exception.

The Northern River Basins Study (Cash, 1995; Northern River Basins Study Board, 1996) was a large-scale application of the weight-of-evidence approach in ERA. The study, a Canadian federal/provincial initiative, strove to understand the cumulative impacts of development on aquatic ecosystems. A cumulative effects assessment was mandated. However, the lack of a generally accepted conceptual framework to integrate the various antagonistic and synergistic effects of multiple stressors on ecosystems was problematic. These multiple stressors included complex industrial effluents, flow regulation, long-range atmospheric transport, and changes to climate that could affect ecosystem components from primary producers to top consumers. The approach to the evaluation of the cumulative effects of these stressors involved the establishment of stressor exposure and the measurement of stressor effects, including the identification of credible cause and effect mechanisms.

The weight-of-evidence approach as used in the risk characterization phase of the ERA paradigm was chosen as the best conceptual tool for the cumulative effects assessment of the basins study. Typical of the weight-of-evidence approach was the synthesis of existing field data across spatial and temporal scales to determine overall pattern to be followed where possible by experimental hypothesis testing to determine causal mechanisms. Criteria were then developed against which field and experimental data could be weighed to determine causality. The ultimate goal was to provide information on the state of the system under study whereby this information was then evaluated and acted on according to regulatory, social, and management requirements (Culp Cash et al., 2000b).

5.2 Generalizations Developed from the Literature Review

The current literature does not provide a readily accessible basis for the construction of generalized conceptual models of multiple-stressor effects on aquatic ecosystems because of the lack of a broader context to multiple-stressor research. However, a few tentative generalizations may be possible based on research to date and extrapolation from broader ecological theory.

The context in which the receptors are chosen is an important factor to consider when examining conceptual models for a risk assessment study. Receptors can be examined at the individual-, community-, or ecosystem-level, with each level showing a differing response to stress. The organizational level that is being studied will determine the type and additivity of effects.

Generally, it appears that stresses of most kinds have a simple additive effect on individual organisms (EIFAC, 1987; Van Der Gaag, 1992), but the response to stress is nonlinear. There is a lower threshold below which no effects are detected, a midregion of predictable dose-response and an upper threshold above which the animals die.

We can anticipate that populations and communities will respond to stress in a predictable, linear fashion (above a lower threshold) because increasingly less-sensitive individuals or species will be removed as the intensity of the stressor(s) increases. Furthermore the effect of multiple stressors is probably more often additive than synergistic or antagonistic at the population level. For example, Rasmussen and Lindegaard (1988) sampled mayflies of the genus *Baetis* from Danish streams containing a very wide range of dissolved iron concentrations. Some *Baetis* species disappeared in streams with iron concentrations as low as 0.2 mg/L, but at least one species was found in all but the most heavily contaminated streams (10 mg/L iron). The gradual decline in representatives of this genus mirrors the stepwise response to be expected from aquatic communities more generally (Leland et al., 1989).

According to Odum (1985), stressed ecosystems maintain processes at the expense of species; consequently stressors may have sharp thresholds of effect, below which the effect is minor or even undetectable (still enough species left to do things). Above this level the effect is abruptly pronounced (no longer enough species or individuals to support essential processes). This has been demonstrated in some streams (Lugthart and Wallace, 1992). If the stressor is stimulatory of growth or production, like heat or nutrients, then a subsidy effect may be observed at low levels (Odum, 1985).

The subsidy side of the equation in Odum's model has been given less consideration than the stress side. Nevertheless, scattered observations in the literature broadly support Odum's conception. Additions of nutrients, organic matter and energy (heat or light) are stimulatory at the species, community and ecosystem levels when the amounts added are small. They tend to become stressors by indirect routes when the levels rise above certain minimum thresholds. For example, increasing levels of nutrients will continue to increase plant production (and hence animal production) until they are no longer limiting to growth. The disruptive effects of these high-nutrient subsidies arise from their unbalancing of competitive relationships, wherein a few aggressive, fast-growing species are favored over the large number of specialized, slow-growing species, and from such indirect effects as smothering, loss of habitat, and deoxygenation during decomposition. The mechanism of stress is fundamentally different from toxic chemicals that produce a stress response directly, at least at the organism level.

Even toxic chemicals may act as subsidies to the ecosystem if the concentration is low enough. Toxic organics contain reduced carbon and therefore are a food source for bacteria; increased microbial production is common in microcosms or natural systems dosed with chronic low doses of toxic organic compounds such as pesticides. Even heavy metals can be stimulatory at low doses. The unexpected stimulation of chironomid populations by metal-contaminated waters at Clark Fork has been mentioned. Members of the chironomid subfamily Orthocladinae often thrive under chronic metals exposures that are toxic to other insects (La Point et al., 1984; Leland et al., 1989; Kiffney and Clements, 1994).

The nature of interactions between stressors and subsidies is not well understood, but it may be possible to predict the combined effect of stressors and subsidies using a simple additive model in which subsidies are given negative weight. In addition to stimulating growth, nutrient or energy subsidies lead to an acceleration of metabolic processes. This acceleration, in turn, can exacerbate or reduce the response to other stressors: fast-growing organisms absorb toxic heavy metals more quickly, for example (Wang and Dei, 2001), but may be more resistant to toxicity because of greater vigor and dilution of toxicants in a larger mass of tissue. All other things being equal, the same effect should apply to subsidies of all kinds.

It is important to distinguish the interactions between the stressors from the interactions between stressor effects. Interactions between stressors are physical and chemical interactions that occur outside the organism and should be predictable from models based on the physical environment. We can envision the effect of an organic toxicant combined with temperature by adjusting the decay rate of the compound, for example. The interaction of stressor effects, however, occurs inside the organism, or in the biotic component of the ecosystem, and can be predicted, perhaps, by systems-based models. Much of the confusion about interactions between broadly different stressors arises from a failure to make this distinction.

Climate change, invasive species, overharvesting, and flow regulation can perhaps be rendered tractable by carefully and painstakingly considering their effects on ecosystems. The risk assessment can concentrate on precisely explicated changes to the system. In the risk assessment framework, it is an issue of problem formulation. For example, instead of trying to study climate change, we can instead use meteorological and hydrological models to predict the effects of climate change, for example a different level and schedule of runoff and stream flow. The risk assessment can then deal with the effects of altered stream flow, a manageable issue, irrespective of its source.

CHAPTER 6.0

THE THEORY OF MULTIPLE-STRESSOR EFFECTS AND A COMPARISON OF EXPERIMENTAL AND OBSERVATIONAL METHODS

6.1 Theory of Multiple-Stressor Effects

Two general types of theory exist regarding the effects of multiple stressors: ecosystem stress theories and mechanistic theories. Ecosystem stress theory focuses on the expected trends in ecosystem structure and function in response to stress. One of the most well-known theories of ecosystem stress is by Odum (1985). This theory was presented in Chapter 1.0, but in general concerns the interplay between stress and subsidy that ecosystems experience in the presence of stressors and how population, community, and ecosystem parameters respond to those stressors. Mechanistic theories, on the other hand, focus on the expected responses of individuals to multiple stressors based on chemical, physiological, and behavioral factors. An example is the Broderius (1991) model for joint toxicity of xenobiotics based on quantitative structure-activity relationships (QSARs). Broderius (1991) suggests that chemicals with a primary and common mode of toxic action can be modeled by a QSAR and that each mode of toxic action should be characterized by a different QSAR. Chemicals within several specific QSARs have been demonstrated to be strictly additive in their joint toxicity. Chemicals with different QSARs generally display a less-than-strictly additive joint toxicity. Broderius states that the majority of industrial organic chemicals display nonspecific or baseline toxicities are strictly additive in their joint acute toxicity. Another example of the mechanistic approach is the classical pharmacological approach as described by Calamari and Alabaster (1980). Extrapolation from mechanistic explanations of multiple-stressor effects to the population-, community-, or ecosystem-level depends on how well the mechanistic theories predict effects on parameters such as birth rate, death rate and recruitment.

What is striking in the literature is how rarely the two general groups of theories are brought together in the characterization of risk from multiple stressors. This gap may be a product of the lack of interaction between the fields of toxicology and ecology. Many authors have described the paradox facing ecotoxicologists (i.e., the single-species data that are the most readily obtained are the least applicable to the community- and ecosystem-level endpoints of most concern). In response to this paradox, toxicologists have been developing increasingly sophisticated tools for landscape-scale assessments of risk [see the special issue of *Environmental Toxicology and Chemistry*, 2000, 19(4)]. During this process, toxicologists have arrived at the same conclusions that ecologists reached some time ago—the primary "multiple stressors of concern" are not the traditional chemical stressors for which so many toxicological tools have been developed. Instead, they are habitat alteration, hydrological modification, overharvesting, invasive exotic species, and global climate change (Harwell and Gentile, 2000).

The challenge for organizations that are concerned about how their activities and discharges contribute to the risk from multiple stressors is that there is a lack of both theory and tools for dealing with the so-called "Class 4" type of multiple-stressor problem. A Class 4 problem is one with multiple classes of stressors (chemical, physical, and biological) from multiple sources with interactions acting on multiple ecological receptors (Harwell and Gentile, 2000). Most of the receiving environments relevant to organizations like WERF qualify as Class 4 problems, often including the priority stressors listed above. It is easy to become discouraged at this point; however, there are theories, tools and approaches that are being used with some success. "Class 4 assessments necessarily involve the statistical and modeling tools of less complex problems but further require weight-of-evidence approaches and, ultimately, expert judgment and adaptive management," according to Harwell and Gentile (2000).

Various tools and approaches will be integral to the study of multiple stressors, notably mesocosms, models, effects-based assessments, and the weight-of-evidence approach. The following section provides a review of mesocosms. A review of advanced models for the assessment of multiple stressors is beyond the scope of this report; however, the review of Moore and Bartell (2000) is recommended as an introduction to the tools currently available and under development. Field-based cumulative effects assessments were reviewed by Munkittrick and McMaster (2000); this report will not repeat that review. The weight-of-evidence approach is fundamental to the characterization of risk from multiple stressors (see Chapter 1.0). Understanding risks from multiple stressors will always require several lines of evidence and considerable professional judgment.

6.2 Mesocosms as a Tool in Multiple-Stressor Research

The experimental approach to the study of multiple stressors, by necessity, must go beyond single-species laboratory methods to multispecies methods using microcosms and mesocosms. The difference between microcosms and mesocosms is arbitrary and largely related to size and application in the laboratory or the field. Field experiments using medium-to-large sized enclosures, artificial streams or artificial ponds are generally called mesocosms. DeNoyelles et al. (1994) define mesocosms as "middle-sized multispecies experimental systems falling between microcosms of volumes less than 10 m^3 and the large, complex, real-world macrocosm of the pond or lake." The focus of the following discussion is on mesocosms.

The rationale for the use of mesocosms is that they provide a link between field observational studies and laboratory toxicity tests by simulating the natural environment while allowing for control of variables, experimental manipulation, statistical replication, and most importantly, providing cause-and-effect results (Dube et al., 1998). Lamberti and Steinman (1993, and chapters therein) reviewed the uses and misuses of one class of mesocosms, artificial streams, in research on stream ecology. They noted a range of applications for artificial streams ranging from ecotoxicology to predator-prey interactions, but cautioned that validation of research results in natural ecosystems is necessary to draw strong inferences.

A compendium of information on mesocosm study design and case histories is presented in Graney et al. (1994). The potential design of mesocosms for ecological research is reviewed below.

6.2.1 Physical Design

Mesocosms come in a wide variety of shapes and sizes. Outdoor artificial stream designs were reviewed by Mitchell (1994). Mitchell compared man-made watercourses with once-through or variable flow regimes, raised troughs through which pumped water is partially recirculated and diversions of natural streams through small, fabricated channels. Mitchell's discussion considered the following elements of system design: water source, flow regime, dosing methods, disposal of effluents, number and size of channels, provision of key habitat characteristics, and assessment of stability.

Water Source

The outdoor artificial streams were usually located near a water source of suitable quality (e.g., an adjacent river, well, bore hole, or lake) or close to an effluent outfall. Water quality parameters routinely monitored were DO, water temperature, pH, total organic carbon, suspended solids, total hardness, conductivity, nitrate, chorine, and phosphate.

Flow Regime

Flow-through systems have three main disadvantages: difficulty in attaining high flow rates, unacceptably high levels of siltation, and the requirement for large quantities of chemicals or effluents to treat multiple streams with accompanying problems of treating large volumes of waste. Recirculation addressed these concerns but required insulation and counter-current heat exchangers to prevent increases in temperature. Furthermore, recirculation may lead to the breakdown of chemicals by microbial activity and may make maintaining stressor concentrations difficult. Mitchell (1994) recommended systems that offer a choice of operating modes to allow flexibility.

Dosing and Effluent Disposal

Accuracy of dosing is a critical feature of mesocosm design. Chemicals are usually delivered continuously from toxicant reservoirs via metering pumps. Measured chemical concentrations rather than nominal concentrations are required. Adequate mixing may have to be ensured by using additional recirculating pumps.

Effluents are usually released directly into an adjacent watercourse or into a wastewater treatment plant.

Number and Size of Channels

The number and size of channels will be determined by a balance between the desire for environmental realism and the constraints of money and time. Mitchell recommends incorporating pool and riffle sections and providing autochthonous and allochthonous inputs even in relatively small systems to provide more environmental realism.

Systems with multiple channels give flexibility of experimental design allowing studies with replicated treatments or an increased number of treatments.

Building Materials

Materials include earth lined with a nonporous, inert, nontoxic material (e.g., bentonite), sealed plywood, stainless steel, and fiberglass. Stainless steel is the most expensive but is robust and easily cleaned between experiments. The choice of building materials is likely to be governed by budget.

Selection of Organisms and Provision of Key Habitat Characteristics

Key habitat features to consider include: selection of appropriate flow regimes, streambed characteristic, water quality parameters, temperature, and light intensity. Provision of pool and riffle habitats to increase the diversity of habitats may be evaluated by conducting pilot studies to determine the relative sensitivities of pool and riffle species assemblages to the stressors. A variety of taxa including bacteria, protozoa, algae, macrophytes, macroinvertebrates, and fish will increase environmental realism. Flora and fauna can be allowed to colonize naturally or can be actively introduced from natural water bodies.

Flow-through systems with continuous recruitment of organisms will self-populate whereas recirculating systems usually require the introduction of populations. It may be necessary to supplement populations of particular interest.

Confirmation that the systems are evenly populated before treatment is required.

Sampling

It is necessary to consider the possible impact of sampling so as to avoid confounding the interpretation of toxicant-related effects. The optimal number and frequency of collection of samples will be the minimum required to allow statistical procedures at the required levels of significance and to fulfill the objectives of the experiment. Inherent within- and between-stream variability will determine the optimum design, but practical limitations will frequently dominate.

Assessment of Stability

Mitchell found that the stability of mesocosms was rarely addressed in the papers he reviewed. Measurements of stability may include the ability to maintain chemical concentrations or the maintenance of ecological diversity and abundance of individuals over time. Data on the performance of control streams can provide information for determination of the degree of effect that it is possible to detect for a particular response.

Mesocosms for lentic (lake) research vary from artificial ponds (indoors or outdoors) to enclosures (often called limnocorrals) within ponds and lakes. DeNoyelles et al. (1994) reviewed 36 studies using various mesocosm surrogates of ponds and lakes. All shared the exposure of a naturally colonized (except for fish) and maintained assemblage of most of the lentic system components. General requirements for pond mesocosms are that they should contain all trophic levels with diversity similar to natural conditions and with all functions of natural systems (Hibachi et al., 1992). Practical constraints are also important, as with stream mesocosms. Differences in size of artificial ponds do not necessarily produce differences in experimental results, as shown by Hibachi et al. (1994) in a study of the effects of a herbicide in 5- m^3 and 15- m^3 ponds. The results were similar in both pond systems. However in a different study on chemical fate and biological impacts of an insecticide, cyfluthrin, in aquatic ecosystems, difference in size influenced aqueous half-life and sediment concentration of the stressor (Philip Johnson et al., 1994).

The U.S. EPA has published requirements for pond mesocosm design in support of pesticide registration (Brassard et al., 1994). The test system has to have 12 or more ponds, each being at least 0.1 acre. As many replicates as possible within each treatment level is suggested. Baseline data on pond construction, characteristics of test water and sediments, and uniformity of biota among mesocosm ponds is required. Supporting data for these include DO, pH, temperature, rainfall, wind direction, and evaporation. Methods for prevention of invasion of extrinsic fish or other biota must be provided and provisions for effluent disposal must comply with federal and state requirements. Water/sediment analysis should include various depths of the water column and sediment and must be conducted throughout the study. Analysis should delineate temporal and spatial heterogeneity of the chemical. Residue analysis of biota must be reported and bioconcentration factors should be estimated. Recommended response variables are: community metabolism and abundance and biomass of phytoplankton, periphyton, filamentous algae, macrophytes, zooplankton, macroinvertebrates, and fish. Taxa richness, species diversity, and community similarities are also suggested variables. Fish length, weight, and growth rate are also required when feasible.

Marine mesocosms range from *in situ* pelagic enclosures to constructed pelagic and sediment systems. Clark and Noles (1994) reviewed marine mesocosms and their potential for testing of pesticides. They point out that marine systems have no small-scale surrogate system on which a physical model can be based. Maximum chemical inputs to marine systems may occur in tidal creeks or marshes located near coastal activities, but the wide diversity in physical scale and processes present in marine systems provide no single conceptual or physical model as being representative. In marshes and tidal creeks, input and resuspension of particulate matter plays a dynamic role in controlling the fate and transport of contaminants, directly altering the bioavailability of chemicals with time. Clark and Noles (1994) suggest that existing marine mesocosm test systems could be adapted to study pesticide effects, but emphasize that such systems should incorporate sediment components with benthic communities. They also caution that such test systems must be designed with some reference system as a model. According to Clark and Noles (1994), there is no consensus on the suitability of, or preference for, physical models of mudflats, tidal creeks or rocky shorelines as mesocosms for pesticide testing. They also point out that expense in construction and design as well as operational costs will limit the numbers of test systems used.

6.2.2 Experimental Design

Three major categories of experimental design exist: hypothesis tests, point estimate tests or hybrid tests that incorporate elements of hypothesis, and point estimate tests (Touart, 1994). Hypothesis tests are used to investigate whether the response measures are different from that of control measures. Hypothesis tests generally consist of a null hypothesis and an alternate hypothesis, such that rejection of the null hypothesis requires acceptance of the alternate. A test of significance provides the justification for rejecting a null hypothesis. Hypothesis tests are used for comparing means and are characterized by having multiple replicates in control and treatment groups. The greater the number of replicates, the more accurately the group variability is defined and the greater the power of the test for detecting differences. Hypothesis tests are best for objectively determining if an effect is real.

Point estimate tests are designed to evaluate regression relationships and estimate an exposure level that will not cause an adverse effect (no observed adverse effects or threshold level) or predict the intensity of an effect at a given exposure level. Regression analysis is used to iteratively fit observed data to a theoretical equation. This analysis requires multiple treatments at various exposure concentrations with an objective of defining an exposure response relationship. Once an exposure-response relationship has been determined, a specified point may be estimated such as an effect threshold level. The greater the number of treatment levels, the greater the confidence in the fitted exposure-concentration line.

Hybrid tests incorporate features of both hypothesis and point estimate tests. Employing multiple replicates and multiple doses once can determine if a given treatment level significantly differs from control, and may estimate how different another treatment level will be above or below the given treatment level. With a limited number of mesocosms, one reduces the number of dose levels to add replicates. Reduced replicates mean reduced power; reduced dose levels mean reduced confidence in estimating the fit of the regression line. A balanced design incorporating, at the least, a control, concentrations representing upper and lower bounds and one intermediate concentration with a minimum of three replicates per treatment has been recommended by the U.S. EPA (Touart, 1994).

The first step in determining the optimal experimental design is deciding the exact goal of the study (Regal and Lozano, 1994). Thus if the goals are specified in a quantified form and a model is specified describing the responses, an optimal design can be applied to find the best design or designs for addressing the question.

The selection of an experimental design must be clearly related to the context for the study. If the test results are within a regulatory compliance context, then the design chosen depends on the regulatory endpoints. For example, the regulatory objectives for testing pesticides under the Federal Insecticide, Fungicide, and Rodenticide Act (7 U.S.C. s/s 136 et seq. [1996]) are that testing must establish the "no unreasonable effect" level, or, in other words, "to negate the resumption of serious risk and to define the intensity and duration of any adverse effects identified" (Touart, 1994). A presumption of risk may be easily negated by demonstration of no effect at worst-case exposure conditions. If detrimental effects are observed, the judgment of whether the risk of these effects is unreasonable becomes very important and may support the use of a hybrid design where effects thresholds can be estimated with confidence and cause/effect relationships can be established.

The results of a mesocosm study are subject to adaptive interpretation by both statistical and nonstatistical analysis. Statistical analysis assists in the objective evaluation of studies and verifies the power of the test to detect an adverse effect. Nonstatistical analysis is performed by professional judgment of qualitative information to ascertain relevance of the statistical results to field conditions.

Pass/fail conclusions are not drawn directly from a mesocosm study. Instead, a weight-of-evidence review is used to place any adverse effects in an appropriate perspective for use in later risk management decisions (Touart, 1994).

A regression design with some replication offers the highest degree of analytical flexibility (i.e., where a concentration-response is obtained, regression can provide a calculated effects concentration with error bars). In the absence of a concentration-response replication, Hurlbert (1984) permits an analysis of variance to be performed and provides values of no observed effects concentration and/or lowest observed effects concentration (Mitchell, 1994).

The nonlinearity in observed concentration-response relationships and large heterogeneity of variance, resulting from variation being inversely proportional to means create statistical challenges when interpreting mesocosm data. Thompson et al. (1994) observed that data transformations are usually inadequate for linearizing concentration-response relationships. They suggest that a three-dimensional nonlinear approach seems most appropriate because it more closely reflects the theoretical curvilinear nature of concentration-response relationships. It also effectively incorporates the dynamic influence of time and eliminates the need for data transformation. Thompson et al. (1994) suggest that modification of the experimental design to increase the number of test concentrations and reduce replication would be required to optimize the nonlinear, response-surface approach to mesocosm studies. They state that some degree of replication in the design is required for estimation of random variation. However, they suggest "disproportionate replication" in favor of fewer test concentrations would offset the effects of inherently high variance for these concentrations. "Applying disproportionate replication and increasing the number of test concentrations would be a cost-effective method of offsetting the variance heterogeneity at the low concentrations and increasing confidence in prediction" (Thompson et al., 1994).

The design of mesocosms can be quite sophisticated, even attempting to simulate the whole ecosystem in some instances. One example of this is the large tank mesocosms at the Marine Ecosystems Research Laboratory (MERL) designed to model Narragansett Bay, an estuarine ecosystem along the northeast coast of the United States (Lambert and Oviatt, 1986 in Klein-MacPhee et al., 1993).

The MERL mesocosms have several advantages: 1) long-term experiments can be conducted and stressors can be added without introduction to the environment; 2) sufficient tanks are available to be able to replicate experimental treatments; 3) a benthic community as well as a water column community can be included; and 4) temperature control and thermal or salinity stratification can be maintained in the water column (Klein-MacPhee et al., 1993). Simulation of the whole ecosystem increases the probability that exposure to stressors occurs through the route that will occur *in situ* and any indirect effects will be observed (Klein-MacPhee et al., 1993).

The usefulness of the MERL mesocosms was demonstrated in studies of several natural stressors and an unusual chemical stressor—a starch blend material being developed as a biodegradable substitute for plastics (Klein-MacPhee et al., 1993). Variables tested included food availability, blooms of a nuisance alga and the presence of benthic organisms. Because of the sophistication of the study design, the authors were able to detect effects on growth rates of larval fish. Fish gut content analysis indicated that feeding on copepods was reduced in the presence of the starch material even though copepods were more abundant in the treated enclosures. They suggested that fish may have been ingesting the less nutritious starch blend particles in preference to copepods, their normal diet. This subtle effect may not have been detected in a simpler test.

La Point and Fairchild (1994) suggest that the focus of mesocosm testing should be on how aquatic environments modify the bioavailability of pesticides. They also state that the measured biotic effects are a function of the duration and frequency of exposure, the size of the organisms exposed, and the trophic status and physical structure of the environment. The magnitude of effect cannot be isolated from the persistence of the chemical in the water relative to the mean time between generations of exposed aquatic organisms. Life history characteristics, particularly body size and generation time, determine the rate of recovery of a population once it has been exposed to a pesticide. However, the long-term results of pesticide exposure also depend on the availability of refugia, food-web relationships (including controls imposed on population size by predators), and behavioral responses to avoid exposure such as drift. "Extrapolating effects from mesocosms to plankton populations in ponds and lakes may be fairly accurate, particularly if the binding potential of the pesticide to organic carbon is known. However, for longer-lived aquatic populations, it will be difficult to extrapolate chronic effects of the pesticide, as measured in the laboratory, to natural systems" (La Point and Fairchild, 1994).

Important interdependent factors in the design and construction of outdoor artificial stream systems ensure that:

- Clear experimental objectives are set;
- required fauna and flora are present;
- representative sampling can be carried out;
- exposure to the test substance can be achieved and controlled in an acceptable way; and
- the system will lend itself to appropriate statistical analysis (Mitchell, 1994).

Systems that allow for a flexible approach to experimental design are more likely to justify large capital and operating costs than systems that are inflexible. The scale of the system should be sufficient to allow experiments to be conducted that satisfy the objectives with the minimum of expenditure. Design and construction characteristics are inevitably constrained by capital and operating costs. This does not imply that a small budget will undermine viable experimentation; on the contrary, it may focus attention more critically on objectives and experimental design.

Some questions regarding mesocosm design are (Mitchell, 1994):

- What constitutes a "successful" outdoor artificial stream experiment?
- Which design factors are of particular significance?
- What are appropriate validity criteria: stability, sensitivity, diversity, etc., and how should they be measured?
- How can the results of studies from different systems be compared and evaluated?

- ♦ Can results from one flowing water environment be extrapolated to others?

DeNoyelles et al. (1994) emphasize the importance of including analysis for indirect effects. They point out that there are very few attempts to describe mathematically the pathways of effects. These pathway analyses may provide measures of the total response of the ecosystem, integrating the array of direct and indirect effects into one response measure, as suggested by Johnson et al. in DeNoyelles et al. (1994). Use of this and other mathematical models to analyze stability within mesocosm ecosystems is a means of applying ecosystem stability theory to mesocosm effects testing.

Two important questions related to study design that have considerable impact on the cost of studies are:

1. What size of mesocosm is required to give useful results for the endpoints identified as important?
2. What level of replication is necessary?

Evidence exists that for some stressors, small pond enclosures or small flow-through artificial streams can provide results that are similar to, or complement, data from larger systems or from field studies (Stephenson, 1994; Dube et al., 1998). Therefore, smaller mesocosms can provide valid and useful data; however, the limitations regarding extrapolation to the field situation must be explicitly addressed.

The level of replication will depend on whether the experimental design is primarily focused on hypothesis testing or on the derivation of concentration-response relationships. It will also depend on the inherent variability of the effects endpoints. Christman et al. (1994) analyzed the precision of various biological response variables (or "metrics") in terms of the percent change that must occur to distinguish treatment effects from natural variability in pond mesocosms. Their results showed that there were considerable differences among individual mesocosms. Taxa richness metrics were from 2% to 40%. Proportion of numerical abundance metrics were from 2% to 60%. Density metrics were from 10% to 100%. Of the 20 metrics evaluated, 7 allowed detection of impact with a 20% change from the mean. Four of these were taxa richness metrics and the other three were proportion of numerical abundance metrics. For all metrics, using six replicates instead of three would usually improve the ability to detect differences between treatments. The authors suggested that a management program to ensure a more even distribution of biota among mesocosms would improve the ability to detect treatment effects. They recommended analyzing pretreatment data to establish which metrics are realistic indicators of effects for a particular study.

6.2.2.1 Endpoints

Endpoints are measures of effects or responses. Some authors define endpoints as conclusions drawn from the results of an experimental study (Touart, 1994). Endpoints can be separated into two categories. The first category may be called "observational endpoints" and includes direct measurements or calculations of specific parameters or phenomena embodied within the test (Touart, 1994). Examples would be a change in numbers of invertebrates or in total biomass of fish. The other category may be called interpretational endpoints; these endpoints are determined indirectly through assessment or judgment of an observational endpoint (Touart, 1994). An interpretational endpoint is explanatory of an observational endpoint. For example, a reduction in fish biomass could be interpreted to indicate poor growth and foretell poor reproductive performance (Touart, 1994).

Fate and effects endpoints that might be included in mesocosm studies are:

- Fate—persistence, physical, chemical, or biological loss processes.
- Organisms—mortality, behavior, growth, and reproduction.
- Populations—abundance and distribution over time for determining the persistence of a population. Change in biomass and density for determining population production. Structure as characterized by size/age class distributions, recruitment, and spatial distribution.
- Communities—Community composition as expressed in species diversity, species dominance, the presence of sensitive and tolerant species and succession. Community organization as measured by functional groups or guild structure and complexity of trophic structure. Community productivity as described by measures of energy or material flow (e.g., turnover rates, nutrient ratios, production/biomass ratios) or the production capacity of functional groups (e.g., the primary productivity of the phytoplankton community).

DeNoyelles et al. (1994) emphasize the importance of including both direct and indirect effects endpoints. They point out that when researchers look for indirect effects, they usually observe some. Indirect effects can include increases in the abundance of organisms because of a release from grazing, predation or competition pressure. Some chemicals can cause changes in physical or chemical conditions. For example, DO declines severe enough to affect fish can result from the death and decomposition of algae and macrophytes killed directly by a chemical. Oxygen increases sometimes occur in response to increased plant growth because of reduced grazing. Concentrations of nutrients can increase following death and decomposition of plants or animals after chemical addition. Altered habitat structure because of the elimination of macrophytes is a common indirect effect of herbicides.

6.2.2.2 Scaling

The issue of scaling is central to experimental studies in ecology as well as to all scientific research (Van Gardinger et al., 1997). Generally, scaling "up" from artificial systems to natural ecosystems is recognized as a major challenge in ecological and toxicological studies. There are several factors that affect the reliability of extrapolation from mesocosm to field. These factors are reviewed below.

Generally, mesocosms are designed to model relevant structural and functional aspects of larger and true ecosystems (Belanger, 1992). To increase the likelihood that stream mesocosms are a good surrogate for the field, physical, chemical, and biological components of mesocosms must be designed with the receiving environment of interest in mind. Crucial physical design considerations for stream mesocosms are size, configuration, water quantity, substrate, temperature, and light (Rodgers et al., 1996).

The size of a model stream mesocosm will determine the range of aquatic life that can be accommodated, the amount of sampling that can be done without causing sampling artifacts, and the longer-term internal sustainability of the test system (Rodgers et al., 1996; Belanger, 1997). Larger test systems allow for the possibility of including fish. The impact of fish in a test system can be very large; therefore, most research is conducted using caged fish (Belanger, 1997). The size of the system must be sufficient to encompass or to capture the effects parameters of interest. However, Belanger (1997) in a metal-analysis of more than 150 model stream mesocosms, found that factors other than size apparently control biological complexity in these systems.

The configuration of a model stream mesocosm can include alternating series of riffles and pools to encompass species assemblages and effects specific to these two habitats. The inclusion of riffles and pools may be particularly important when the primary test endpoints are for benthic invertebrates.

Flow in model stream mesocosms must be calibrated to field conditions as much as logistics allow. This may include incorporating variability in flow, depending on the hypotheses being tested. Flow velocities and volumes can be problematic, particularly if the source water contains an abundance of suspended solids at relatively high flows that cannot be replicated in the mesocosm (Rodgers et al., 1996). In this case, severe sedimentation in the mesocosm may occur.

The substrate that is selected for mesocosms can influence chemical behavior and biological responses; therefore, if hypotheses being tested include the role of sediment, care must be taken to include the range of sediment particle size and other factors (e.g., organic content) of interest in the study design. This may be of particular importance when benthic invertebrate responses are the primary measure. Differences in substrate may not be as important for other biologic indicators. For example, periphytic community structure and function were highly similar in replicate stream mesocosms on both tile and cobble substrata (Lowe et al., 1996). In other cases, the experiments may be conducted without sediments or using sediments of uniform particle size.

Temperature is a fundamental factor influencing chemical fate, mode of toxic action, and biologic response. When receiving environments are subject to pronounced seasonal temperature differences, inclusion of temperature as a variable in the design may be required. Alternatively, the temperature conditions expected to correspond with maximum response may be used. Mesocosms can be effective tools in examining the effects of seasonally varying stream water temperature on processes such as biodegradation and subsequent toxicity. For example, the effects of decreasing water temperature on surfactant biodegradation and algal sensitivity were evaluated using a series of four mesocosm studies where temperature were from $0^{o}C$ to $28^{o}C$ over all studies and temperature declines were $9^{o}C$ to $14^{o}C$ over the course of each individual study (Lee et al., 1997).

Light quantity and quality are also crucial in mesocosm trajectory and development (Rodgers et al., 1996). Many streams are characterized by some shading; however, there are also many streams with open canopies that receive full sunlight. Light can directly affect chemical fate (via photodegradation or photoactivation processes). Light determines primary productivity. Many taxa respond directly to light either by avoidance or attraction. Therefore, light conditions (and the seasonality of light conditions) must be considered in mesocosm design.

Crucial chemical considerations in mesocosm design relative to scaling include water quality. Studies that used well water as a dilution source can produce different (and less scalable) results than studies that used upstream surface water as a dilution source. The reasons for this can include differences in important modifying factors such as hardness, pH, alkalinity, and organic ligand content. In addition, well water has an inherently small number of biological propagules; therefore, there is less continuous colonization within the mesocosms (Belanger, 1997).

Crucial biological considerations in mesocosm design include biota sources or propagule sources, the diversity of aquatic life represented in the mesocosm, the number of trophic interactions and pathways present, and the degree to which nutrient cycling and energy flow are represented. In addition, it is essential that model ecosystems are in approximately the same ecological state before chemical exposure (Lowe et al., 1996).

Model stream mesocosm biota sources or propagule sources can range from gnotobiotic studies through to studies where colonization and establishment of biota are assisted by sediment transfer and organism transfer. Some mesocosm studies rely on the source water, which contains propagules such as eggs and larval organisms that may colonize and inhabit the mesocosms. Frequently, hybrid protocols are used in which mesocosms are initially colonized with assistance from sediments with extant biota as well as additionally colonized from propagule sources in the source water (Rodgers et al., 1996).

A diverse assemblage of aquatic life in mesocosms is desirable in order that a wide range of organism sensitivities can be evaluated and multiple trophic interactions are represented. In many instances, it may be desirable to include fish, especially if fish are particularly sensitive to the test chemical (Belanger, 1992). The presence of macrophytes in mesocosms can have a confounding effect, especially during growth or decay phases (Blake, 1994, in Belanger, 1997). Epiphytic algal growth on macrophytes can be a substantial contributor to increased species diversity (Lehtinen et al., 1991, in Belanger, 1997). Fish can have important effects via size-selective predation. The impact of fish can be so large that many researchers use caged fish in mesocosms.

Mesocosms should reflect essential elements of ecosystem function (nutrient cycling and energy flow) (Belanger, 1992). This may include primary and secondary productivity measurements, as well as total respiration.

Over time, mesocosms have evolved that emphasize flow-through exposures, use of natural sources waters and regression-based experimental designs (Lee et al., 1997, Belanger, 1995, Dyer and Belanger, 1999, Belanger et al., 2000, 2002). These designs produce a high degree of biological complexity and reasonable levels of variability for ecologically relevant endpoints. Mesocosm size appears to be less important in controlling biological complexity (Lowe et al., 1996, Guckert, 1996).

In summary, if mesocosms are designed with careful consideration of the physical, chemical and biological factors affecting mesocosm performance and relevance to the receiving environment, the issues with scaling up to the field should be manageable and the mesocosm data can be used with confidence in assessment of risks. Direct comparisons of mesocosm and field data are rare; however, a comprehensive comparison of the Procter and Gamble experimental stream facility to routine field samples collected from a range of streams across Ohio showed that the facility was sensitive and serves as a good surrogate for the field (Dyer and Belanger, 1999). In another comparison of fate and effects measured in laboratory, mesocosm, and field systems, Belanger et al. (2000) found that there was good correspondence among the three lines of evidence regarding responses of aquatic communities to alcohol ethoxylate.

6.2.2.3 Advantages and Disadvantages of Mesocosms

A number of recognized advantages can be realized in mesocosm studies. Mesocosms allow for true treatment replication and a way to determine dose-response concentrations to be chosen appropriately (Fischer, 1994). Effects on the test organisms are usually readily apparent and statistical analyses are relatively straightforward (Lamberti and Steinman, 1993). Mesocosms provide the setting to detect both direct and indirect effects. For example, direct effects can develop into detectable indirect effects on other organisms or on progressively higher levels of organization such as populations, communities and the ecosystem (DeNoyelles et al., 1994). Mesocosm studies also allow for treatment replication and hypothesis testing.

Model or experimental stream ecosystems have been used as research tools for decades. More recently, experimental stream ecosystems have been used to provide refined and comprehensive data sets for hazard and risk assessment purposes. The results of model ecosystems studies are considered for the final tier in regulatory assessments (U.S. EPA, 1984; EEC, 1994; Zeeman, 1995).

The Procter and Gamble experimental stream facility program is a good example of such systems being used to evaluate the risk of various consumer product ingredients, especially surfactants (Belanger, 1992, Belanger et al., 1995, 2000, 2002; Dyer and Belanger, 1999).

Mesocosm studies on integration of fate and ecological responses by Belanger et al. (2002) on linear alkyl benzene sulfonate and 25-6 alcohol ethoxylate (Belanger et al., 2000) indicate the advantage of using mesocosms that have a robust experimental and physical design. These studies further demonstrate the heterogeneity, diversity, and sensitivity of an aquatic community established within the mesocosms.

The advantages of mesocosm testing over laboratory tests or single-species toxicity tests, especially in evaluating fate, effects, and exposure include (Belanger, 1992):

- Single-species tests cannot assess biological interactions such as predation and competition;
- Single-species tests are unable to evaluate the fundamental ecosystem process such as nutrient cycling or spiraling and energy flow;
- The chosen single species used in risk assessment may not represent an important link in a functioning ecosystem;

- ♦ Mesocosms can appropriately evaluate key elements of system assimilation capacity such as influence of water quality (e.g., organic carbon, alkalinity, fluctuating environmental conditions) on test chemicals availability; and
- ♦ Mesocosms can be used to determine ecosystem structural and functional redundancy.

Test systems have now evolved that emphasize flow-through exposures, use of natural source waters, and regression style experimental designs (Belanger, 1997). These factors collectively reduce problems associated with pseudoreplication. More and more studies are indicating that the numbers of test system assigned to treatments (replicates) should be determined based on the study needs in light of scientific and regulatory considerations (Belanger, 1997). A very common approach employed by ecotoxicologists using model ecosystems is to maximize treatments and minimize replication (Kosinski, 1989). Hurlbert (1984) argued that much of the field ecology literature is filled with "pseudoreplication," where researchers have inappropriately ignored assessment of treatment variability. The regression approach used in ecotoxicity test designs may be the most appropriate when model ecosystems are used in a risk assessment context.

Several model ecosystem studies that were compared with natural field-based communities demonstrated high degree of correspondence for biological attributes. Well-designed model stream ecosystems have a high degree of biological complexity attributes and reasonable levels of variability for ecologically relevant endpoints (Belanger, 1997).

Another advantage of performing mesocosm studies is the ability to use natural surface waters to present the test material in a more environmentally relevant and realistic form than in the laboratory (Crossland et al., 1992, Belanger et al., 2002).

Several potential disadvantages must also be considered when conducting mesocosm studies of stressors. The quantity of experimental contaminants to be added in mesocosms depends on theoretical predictions of environmental situations. The dosing of predicted concentrations occurs evenly over the entire surface, unlike the natural situation where uneven mixing of experimental contaminants. There is only a limited possibility for a repopulation by receptors. The presence of predators in mesocosms creates additional logistical difficulties such as obscuring the effects on receptor mortality caused by experimental changes. Mesocosm systems have not reached a climax, but are at various stages of succession. Not all elements of an ecosystem are present in mesocosms (Fischer, 1994). Interpretation of statistical differences requires care. For example, in a study of the effects of endosulfan in mesocosms and farm ponds, the only major effect clearly attributable to endosulfan was localized fish kill around shallow areas at runoff locations. All other statistical differences were attributable to either natural sources or "typical" biological variation (Fischer, 1994).

Both mesocosm and field studies have common problems of interpretation, representativeness, and extrapolation. Questions of interpretation include: How should one interpret declines occurring earlier or later than in control systems? Are temporary differences ecologically adverse effects under all circumstances? Are the causes of the observed effects unequivocally direct or indirect results of the exposure to the stressor? How representative are the results for other types of aquatic systems? How do we extrapolate the results to other environmental conditions? (Fischer, 1994).

Some authors have been critical of the use of constructed species assemblages for use in mesocosm work because they involve testing of groups of aquatic organisms simultaneously, with little concern for natural species interactions. "These assemblages are often highly synthetic and are frequently selected out of convenience and the need for standardization rather than ecological relevance" (Clements et al., 1988). The use of natural benthic invertebrate species assemblages that are allowed to colonize on the experimental substrate partially addresses this concern. Clements et al. (1988) used such an approach in their study of copper and zinc effects on benthic invertebrates in outdoor experimental streams. They used trays colonized by the natural benthic assemblage placed in experimental flow-through streams supplied with water from an adjacent river. The main advantage of this approach is that it allows the observation of community-level responses to stressors. Furthermore, it can be more environmentally realistic because detritus and periphyton are collected together with the benthic invertebrates, thus providing a natural food source. Finally, this approach provides an opportunity to observe responses of organisms that are rarely included in standard laboratory tests.

Stephenson (1994) questioned the value of observations of indirect effects in mesocosms as a means of predicting indirect effects in other systems. They discussed pond mesocosm studies of pesticide effects where there was an increase in diaptomid zooplankton populations was observed. It was not possible to determine if the increase was a response to increased food availability following the removal of competitors (daphnids), or if it was a response to the removal of predators, killed by the chemical treatment. In another study on fish growth, reduced growth of rainbow trout was caused by the death of invertebrates that formed an important element of their food base. However, it was not clear whether such effects would be observed in ecosystems where a greater variety of food organisms was present. Also, Stephenson pointed out that if a different species of fish with different feeding habits had been present, the results may have been quite different. Stephenson uses these examples to draw the following conclusion: although mesocosm studies enable secondary or indirect effects to be observed, it is not likely that this knowledge will enable accurate prediction of the likelihood of such effects occurring in other aquatic systems. This will only occur if there is a detailed understanding of the mechanisms by which the effects came about and an equally detailed knowledge of the ecosystems for which predictions are to be made (Stephenson, 1994).

CHAPTER 7.0

GAPS IN THE MULTIPLE-STRESSOR LITERATURE

7.1 Gaps in the Theoretical Basis for Multiple-Stressor Effects

7.1.1 Overview

A rigorous, theoretical basis for the prediction of multiple-stressor effects does not exist. Despite the wealth of observational data, the existence of several useful tools for interpretation of cause/effect relationships (including formal SI methods), and the studies reviewed in this report, there are no tools that allow a confident, *a priori,* prediction of ecosystem response to multiple stressors. Retrospective studies can use weight-of-evidence approaches and the standard rules for establishing cause/effect relationships to piece together a plausible story regarding the stressors most likely to have caused the observed responses. However, these retrospective studies are time-consuming, data-intensive, and tend to result in analyses that cannot be readily extrapolated to other cases. Furthermore, uncertainty associated with these retrospective studies still centers around model error (i.e., our imperfect understanding of how multiple stressors can be expected to interact and how ecosystems can be expected to respond). Instead, we default to a standard assumption of additivity (in the case of chemical stressors) or to a case-by-case, professional judgment process.

The ultimate goal for the assessment of risks from multiple stressors is to have an understanding of multiple-stressor effects that is sufficiently robust to support decision making. In other words, we aspire to combine observational and experimental evidence for pattern and process responses to multiple stressors in a manner that allows confident analysis and prediction. The primary outcomes would be credible, defensible, and practical conceptual models that can be used as the starting point for risk assessments at a wide variety of sites.

A common finding throughout the multiple-stressor literature is that most articles are an inventory of stressors with no study of the interactions between the stressors. The interactive effects of these stressors are not addressed, largely because it is difficult to apportion the effects observed to the variety of stressors present. Studies that use the weight-of-evidence case for cause and effect relationships among stressors and receptors make an attempt to address these interactive effects.

There is considerable difficulty in quantifying the effects of multiple stressors that interact in a nonlinear manner. With the benefit of extensive research on specific interactions, we can produce reasonable predictions of cumulative stress (e.g., from many years of study of aluminum plus pH). It is a greater challenge to calculate the cumulative stress from two stressors whose interaction has not been thoroughly studied.

The response of ecological communities and ecosystems to multiple stressors provided by authors in the current literature is usually focused on a specific site or a particular combination of stressors. There is rarely any reference to a more generalized model of multiple-stressor effects apart from the discussion of the expectations regarding additivity versus synergism. Many multiple-stressor studies are merely an inventory of the various stresses and upsets affecting the ecosystem, without attempting to assess the interactions among them.

There are a few studies that define the problem in terms of building a weight-of-evidence case for cause/effect relationships among stressors and receptors. Some studies also refer to a suggested general response pattern to stressors (e.g., Munkittrick and Dixon, 1989; Gibbons and Munkittrick, 1994). However, these papers seldom place the definition of the problem in a wider framework provided by a general model such as Odum's model of ecosystem stress (Odum, 1985).

There is a lack of a broad context approach to multiple-stressor research, therefore, the current literature does not provide a readily accessible basis for the construction of generalized conceptual models of multiple-stressor effects on aquatic ecosystems. Based on research to date and extrapolation from broader ecological theory, it may be possible to form a few tentative generalizations. These generalizations are presented in Section 5.2.

7.2 Data Gaps in the Context of Risk Assessment

7.2.1 Types of Stressor Combinations Tested

There are surprisingly few true multiple-stressor experimental studies (Table 7-1). The largest number of multiple-stressor studies is for combinations of metals and metals plus other stressors. A large volume of literature was available on the interactions of metals and pH, but fewer studies discussed the interactions of metals and organic substances such as PAHs, PCBs, and pesticides. Studies of pesticide combinations were very rare in the literature. Studies of pesticides with other stressors were more common than studies just involving multiple pesticides, but experimental studies of combinations of organic compounds other than pesticides were very rare. Most organic compound multiple-stressor studies involved PCBs. Very few studies involved PAHs, except as uncharacterized mixtures.

Table 7-1. Number of Experimental Multiple-Stressor Studies by Stressor Category.

Stressor Category	Number of Studies
Metals	30
Metals plus other stressors	20
Pesticides	7
Pesticides plus other stressors	12
Organic compounds plus other stressors	23
Nutrients/major ions plus other stressors	26
Habitat change/invasion of nonindigenous species	6

There is a notable lack of experimental studies of combinations of stressors associated with agriculture, except for uncharacterized mixtures such as "agricultural pollution."

Few studies have been found on the effects of the "top four" stressors identified by the U.S. EPA Integrated Risk Project (Harwell and Gentile, 2000). These four stressors are climate change, habitat alteration or removal, flow regulation, and invasion of exotic species. It would be possible to design experiments mimicking the increased temperature, reduction in precipitation and increased frequency and severity of extreme weather events associated with climate change in mesocosms or artificial streams, in which the effects on the biota would be measured. To our knowledge this has not yet been done for any climate change scenarios, let alone in combination with other stressors. Extrapolation from mesocosms to global effects, however, will prove to be difficult.

7.2.2 Receptor Selection

Although the importance of building a relationship between stressors and receptors is of great importance, so too is the correct choice of receptors to study. The usefulness (or lack thereof) of the selected receptors as suitable indicators of the community or ecosystem is seldom critically examined in the literature.

The majority of experimental studies discussed single-species receptors, or interactions between two species under the presence of stressors. Few studies examined multispecies receptors, unless they were community assemblages of aufwuchs. Whole community studies were very rare, likely because of the level of complexity inherent in larger systems.

Metal studies focused mostly on fish species. Papers studying the effects of multiple stressors on periphyton were noticeably absent. Studies of the interactions of metals and other stressors showed a more even division between fish, benthic invertebrates, and algae, although aquatic macrophyte studies were underrepresented in the literature.

Few studies provided an explanation of why particular receptors were chosen. Receptors were sometimes taken from the study areas to the laboratory or field mesocosm to represent the receptors found in the study lake, river, or stream

7.2.3 Exposure Assessment

The measures of exposure in aquatic systems covered in the current literature mostly include the traditional measures of concentrations in water and/or sediments and are sometimes supplemented with measures of tissue residues.

There is a need for studies relating tissue residues to effects. Measurement of stressor concentration in tissue would provide an estimate of "dose" and would reduce the uncertainty associated with estimates of bioavailability.

There is also a need for a wide-ranging exercise that produces hypothesized critical pathways in the presence of the priority stressors. This exercise may reveal unique critical pathways that would not have been thought about before.

When multiple-stressor studies combine chemical and nonchemical stressors, the cumulative exposure to these stressors is difficult to describe. The studies to date tend to be a simple listing of the chemical exposure (as concentration or tissue residue) and the relative changes in nonchemical stressors (such as relative change in habitat). True cumulative exposure measures require investigation.

7.2.4 Effects Measures

No unique measures of effects that would contribute to our understanding of the mechanisms underlying the toxicology of metal mixtures were observed in the literature. Measures of effects on individuals in experimental studies are either standardized single-species toxicity test measures or measures specific to the toxic mechanism.

Effects measures relevant to populations (e.g., birth rate, recruitment, mortality rate) require development. The sensitivity of community-level measures and the usefulness of these measures in testing hypotheses regarding multiple-stressor interactions require more critical evaluation.

Measures useful in distinguishing additive, antagonistic, or synergistic effects require identification and development. Traditional single-organism or community measures may not be the most informative regarding multiple-stressor effects because of high variability, or the lack of any connection with the mechanisms of multiple-stressor interactions.

7.2.5 Assessment Endpoints and Conceptual Models

There are no examples of assessment endpoints for experimental studies of multiple stressors in the literature. Conceptual models are also missing, except for the occasional mention of the expectation of additive versus synergistic effects.

Assessment endpoints must be measurable ecosystem characteristics that adequately represent management goals (U.S. EPA, 1998). For example, if the management goal is to reduce the risk to striped bass populations from multiple-stressor interactions to acceptable levels, then the assessment endpoint must be sensitive to the identified stressors, yet respond in different ways to the different stressors. Striped bass populations may be adversely affected at several life stages, in different habitats, through different ways and by different stressors. A useful assessment endpoint would stipulate the attributes of striped bass populations most likely to respond to the particular combination of stressors under investigation. This assessment endpoint would be accompanied by a set of measures that encompass the range of responses of the various life stages to the different stressors.

Conceptual models consist of two main parts, a set of risk hypotheses that describe predicted relationships among stressor, exposure and assessment endpoint response, and a diagram that illustrates the relationships in the risk hypotheses (U.S. EPA, 1998). Risk hypotheses are specific assumptions about potential risk to assessment endpoints and may be based on theory and logic, empirical data, mathematical models or probabilistic models (U.S. EPA, 1998). There was a notable lack of risk hypotheses meeting these requirements in the literature. Conceptual model diagrams are a visual representation of these risk hypotheses. There was also a notable lack of conceptual model diagrams in the literature.

The lack of multiple-stressor conceptual models is a primary source of uncertainty in the assessment of multiple-stressor risks. Because of this lack, the risk characterization of multiple stressors may misinterpret the actual risks. Therefore, any contribution to the development of general conceptual models for multiple-stressor situations will be very valuable and was a particular focus of the accompanying report to this literature review.

7.3 The Need for an Experimental Approach

The generation of testable hypotheses is probably the most useful aspect of observational research in multiple stressors. However, observational studies are not sufficient to provide a sound theoretical basis for understanding multiple-stressor effects in aquatic ecosystems. Although observational studies have been useful in identifying the general attributes of systems exposed to multiple stressors, the actual role of particular stressors versus the role of the mixture of stressors is difficult to deduce from these data. Furthermore, the ecological processes leading to the observed responses can only be speculated on. Thus, the ability to predict the response of, for example, a periphyton or benthic invertebrate community is limited. For example, a very extensive field database from the Clark Fork River could not be used to distinguish whether the effects of sediment metals was additive, dominated by one metal (such as copper), or were a complex function of the particular metal mixture, and the specific physical and chemical characteristics of the Clark Fork River.

An experimental approach is required to test hypotheses arising from conceptual models of multiple-stressor effects. Experimental results can be used in combination with observational field studies in an iterative manner to produce increasingly robust conceptual models. The companion study design document (WERF Report 00-ECO-2A) includes provision for complementary field studies; however, its emphasis is on the experimental approach because of the demonstrated lack of a sound experimental basis for multiple-stressor theories.

APPENDIX A

LIST OF DATABASES SEARCHED

APILIT. 1965-Oct 2000. API EnCompass, New York, NY.

Aquatic Sciences and Fish Abstracts. 1978-Oct 2000, Cambridge Scientific Abstracts, Bethesda, MD.

Biosis Previews. 1969-Oct 2001. BIOSIS, Philadelphia, PA.

CAB Abstracts. 1972-Sep 2000. CAB International, Wallingford, Oxon, U.K.

Conference Papers Index. 1973-Sep 2000. Cambridge Scientific Abstracts, Bethesda, MD.

Dissertation Abstracts Online. 1861-Jul 2000. UMI/Bell & Howell, Ann Arbor, MI.

Ei Compendex. 1970-Oct 2000. Engineering Information, Inc., Hoboken, NJ.

Energy Science and Technology. 1974-Oct 2000. U.S. Department of Energy, Oak Ridge, TN.

Enviroline. 1975-Oct 2000. Congressional Information Service, Bethesda, MD.

Environmental Bibliography. 1974-Jul 2000. Environmental Studies Institute, Santa Barbara, CA.

FEDRIP: Federal Research in Progress. Oct 2000. National Technical Information Service, Springfield, VA.

FLUIDEX: Fluid Engineering Abstracts. 1972-Sep 2000. Elsevier Science, SPD, New York, NY.

General Science Abstracts/Full-Text. 1984-Sep 2000. The HW Wilson Company, Bronx, NY.

GeoArchive. 1974-Sep 2000. Geosystems, Oxon, U.K.

GEOBASE. 1980-Oct 2000. Elsevier Science, SPD America, New York, NY.

GeoRef. 1785-2000. American Geological Institute, Alexandria, VA.

Inside Conferences. 1993-Oct 2000. The British Library Document Supply Centre, Wetherby, West Yorkshire, U.K.

JICST-Eplus. 1985-2000. Japan Science and Tech Corporation, Information Center for Science and Technology, Tokyo, Japan.

Lewis WM Jr. 1999. Investigations of the relationship between biodiversity and ecosystem function by examination. NSF Dissertation Research, University of Colorado at Boulder, CIRES (Coop. Inst. for Research in Environmental Sciences).

McGraw-Hill Companies Publications Online. 1985-Oct 2000. The McGraw-Hill Companies Publications Online, Inc., Hightstown, NJ.

Meteorological and Geoastrophysical Abstracts. 1970-Nov 2000. American Meteorological Society, Boston, MA.

NTIS. 1964-Nov 2000. National Technical Information Service, U.S. Department of Commerce, Springfield, VA.

Oceanic Abstracts. 1964-Nov 2000. Cambridge Scientific Abstracts, Bethesda, MD.

Pascal. 1973-Oct 2000. Scientific and Technical Information Institute of the Centre National de la Recherche Scientifique, Vandoeuvre-les-Nancy, France.

Pollution Abstracts. 1970-Nov 2000. Cambridge Scientific Abstracts, Bethesda, MD.

Science. 1996-Jul 1999. American Association for the Advancement of Science, Washington, D.C.

TULSA (Petroleum Abstracts). 1965-Oct 2000. Petroleum Abstracts, The University of Tulsa, Tulsa, OK.

WasteInfo. 1974-2000. Waste Management Information Bureau, Abingdon, Oxon, U.K.

WATERNET. 1971-Q3 2000. American Water Works Association, Denver, CO.

Water Resources Abstracts. 1967-Sep 2000. Cambridge Scientific Abstracts, Bethesda, MD.

Wilson Applied Science and Technology Abstracts. 1983-Sep 2000. The HW Wilson Co., Bronx, NY.

Wilson Biological & Agricultural Index. 1983-Sep 2000. The HW Wilson Co., Bronx, N.Y.

Glossary of Terms

Additive	Each substance contributes to the same toxic effect in the same receptor by the same mechanism, with the observed toxicity equal to that expected by adding the toxicities of the individual agents.
Antagonism and Synergism	"A consensus is lacking in the literature regarding the operational definitions of synergism and antagonism among multiple stressors" (Folt et al., 1999). We present the definitions associated with three models of the combined effects of multiple stressors by Folt et al. (1999). The literature was reviewed with this understanding of the various models in mind. *Simple Comparative Effects Model* : The effect of stressors in combination is equal to the effect of the single worst or dominant stressor (Bruland et al., 1991). This is somewhat analogous to Liebig's law of the minimum where only the nutrient that is most limiting affects growth. Similarly, when the worst stressor is present, lesser stressors have no additional impact. *Additive Effects Model*: Synergism or antagonisms occur when the combined effect of multiple stressors is greater (synergism) or less than (antagonism) the sum of effects elicited by individual stressors (Hay et al., 1994 in Folt et al., 1999; Hay 1996 in Folt et al., 1999). This is the most common scientific use of these terms. The additive model has been used frequently on studies of pesticide or metal mixtures. *Multiplicative Effects:* The combined effect of multiple stressors is equal to the product of individual effects. This model has been used to describe a number of competitive interactions and is applicable when stress from one source can be further operated on probabilistically by another source. For example, if low light reduces yield to 55% of optimal and low nitrogen reduces yield to 45% of optimal, combined effects of low light plus low nitrogen would be predicted to be (0.45 x 0.55 ~ 0.25) or 25% of optimal yield (Folt et al., 1999).
Body Burden	The amount of a substance in the body that is either being metabolized, stored or excreted.
Chemical Concentration	Quantifiable amount of a chemical in environmental media.
Chemical of Concern	A chemical in environmental media that is being studied for its potential presence and impact on a receptor.
Conceptual Model	A qualitative or descriptive statement of hypotheses concerning the nature of the potential causal relationships between human activities, the resulting natural and anthropogenic stressors, and their impacts on human and ecological systems (U.S. EPA ,1992, 1998). Conceptual models are often explained by using schematics that illustrate causal relationships. Conceptual models generate "impact hypotheses" that guide the collection and interpretation of data (Harwell and Gentile, 2000). The conceptual model can be revised at each stage of an environmental risk assessment.
Contaminant	A chemical in the environmental media that has either been introduced by human activity or concentrated to levels not naturally found in the area, so that adverse effects may be occurring.

Critical Effect Size	An estimate of the amount of change required to be biologically meaningful—at the population-, community-, or ecosystem-level.
Cumulative	The combination of effects from different activities or sources in an area that has confounding effects on an ecosystem which may not be visible at the individual level.
Cumulative Effects	The impact on the environment that results from the incremental impact of a human activity when added to other past, present, and reasonably foreseeable future activities (Harwell and Gentile 2000). This definition could refer to a single-stressor (such as BOD) that increases incrementally with each additional discharge point to a river. However, this definition is broad enough to include the action of multiple stressors within the same timeframe and same spatial context, and also across time periods. With this broad definition in mind, it can be concluded that cumulative risk assessments and cumulative impact assessments are two different approaches used to assess cumulative effects.
Dose	A measure of integral exposure. Examples include the amount of a chemical ingested, the amount of a chemical actually taken up and the product of the exposure concentration and the duration of the exposure.
EC 50	Effects concentration. The concentration of a chemical or stressor in the water that results in observable effects in 50% of the test species in a given time.
Ecological Endpoint or Assessment Endpoint	An explicit expression of valued environmental attributes that can be defined in operational terms and are amenable to measurement or prediction (Harwell and Gentile, 2000). This definition includes both ecological importance and/or relevance and societal value. Selection of ecological endpoints should result from dialogue between the risk assessor and risk manager during the initial phase of a risk assessment and should reflect stakeholder concerns (Harwell and Gentile, 2000). The term "assessment endpoint" is nearly synonymous with the term "valued ecological component (VEC)" used in the impact assessment literature, although a VEC can also simply be a particular species (which would then be synonymous with "receptor").
Ecologically Significant	The selection of assessment endpoints requires an understanding of what constitutes ecological significance. Harwell and Gentile (2000) suggest that ecological significance relates to determining: ♦ whether a chance that is detected or projected in the ecological system of concern constitutes a change of importance to the structure, function and/or health of the system and exceeds a variance estimate (i.e., placed into the content of natural variability); and ♦ whether such a change in the ecological system is of sufficient type, intensity, extent and/or duration to be important to society.
Effects	A noticeable change in the receptor beyond normal variability caused by a chemical of concern or other stressor.
Exposure	The contact reaction between a chemical and biological system, or an organism.
Exposure Concentration	The concentration of a chemical in its transport or carrier medium at the point of contact.
Fate and Transport	The movement and final location of a chemical through different media. May include partitioning, uptake, and degradation.
Gnotobiotic	A community that only contains known organisms or species (Taub, 1969)
Impacts	A noticeable change in the receptor in reaction to a stressor. See also "Effects."

Indicator	A characteristic of the environment that provides evidence of the occurrence or magnitude of exposure or effects. Formal expressions of the results of measuring an indicator are referred to as measurement endpoints. Abundance, yield, and age/weight ratios are indicators of population production. The presence of female-like characteristics in male fish is an indicator of endocrine system disruption.
Indicator Species	A species that is surveyed or sampled for analysis as a representative of the biotic community, some functional or taxonomic group, or some population that cannot be readily sampled or surveyed.
LC50	Lethal concentration. The concentration of a chemical or stressor in the water that kills 50% of the test species in a given time.
Linkages	The pathways within a conceptual model that connect the stressor with the pathway, receptor and endpoint.
Loadings	The discharges to a system from various different sources. Loadings may be industrial or municipal, spills, chemical application, leaching from groundwater, atmospheric deposition, etc. Loadings are often expressed as the quantity of chemicals discharged (or loaded) to a system.
Measures of ecosystem and receptor characteristics	These are environmental variables that are not identified as stressors of concern, but that may also cause the measured effects to change. These characteristics describe attributes of systems and the organisms of concern that may affect the relationship between the measured effects and exposure to the stressors of concern. Included in this category are life-history characteristics of the species of concern, ecosystem characteristics that affect the behavior and location of these species, and attributes of the system that affect the movement, intensity, and availability of the stressors. Measures of ecosystem and receptor characteristics also are synonymous with the statistical terms "independent variable" or "covariate" (Fairbrother and Bennett 2000).
Measure of Effect	Variables that respond to stressors. Synonymous with the statistical terms "dependent variable" or "response variable" (Fairbrother and Bennett 2000).
Measure of Exposure	Quantification of stressors of concern. Synonymous with the statistical term "independent variable"; also synonymous with "treatment" if the study is experimental (Fairbrother and Bennett, 2000). *Fairbrother and Bennett (2000) note that in risk assessment, the differences between a measure of effect, a measure of exposure and measures of ecosystem and receptor characteristics generally are a matter of perspective and the way the environmental manager frames the particular question. What may be perceived as an attribute of concern and a measure of effect in one system may become a measure of exposure in another system if it also affects another attribute of greater concern. It is more important to state clearly the questions that are being addressed: what environmental attributes are of concern, what potential stressors are being considered, and what additional environmental variables are present and may change the impact of the stressors?*
Mesocosm	A composite physical and biological model of an ecosystem, intermediate in scale between a microcosm and macrocosm, with a level of organization as similar as possible to the natural world.
Multiple	Two or more.

Multiple Stressors	The combined effects of multiple human influences on populations and ecosystems. These commonly include conventional water pollution, toxic chemicals, land-use change, harvesting, and many other qualitatively different disturbances (Barnthouse et al., 2000). Two or more stressors acting simultaneously.
Pathways	The means through which a chemical or other stressor comes in contact with and affects the receptor. For example, at a site with lead-contaminated soil, the pathways for intake of lead in humans are ingestion (soil on foods, children eating soil), inhalation of dust, and direct absorption through the skin.
Propagule	The dispersal stage of a plant or animal that will give rise to new growth.
Receptor	The population, community, organism, organ, or target tissue that is affected by the stressor.
Reference Condition Assessment	The reference condition is defined as "the condition that is representative of a group of minimally disturbed sites organized by selected physical, chemical, and biological characteristics."
Species Tolerances	The physiological and ecological tolerances and requirements describe the hardiness of a species. The more hardy a species is, the greater its ability to adapt to quickly changing environments (i.e., more tolerant)
Stress	An external influence on a population, community, organism, organ, or target tissue that causes an affect, either positive or negative.
Stressor	The chemical, physical, biotic or environmental factor that alters an ecosystem beyond the range of normal variation or from its natural state (Breitburg et al., 1999).
Stressor of concern	A factor identified by the investigator as being outside the normal range of concentration occurrence, usually because it exceeds background or baseline concentrations, levels or frequency, or because it exceeds water or sediment quality criteria. The stressors of concern are a subset of the total stressors acting on an ecosystem.
Synergistic	Two or more toxic substances interact such that the toxicity of the mixture is greater than would be expected if the substances were acting additively or independently. For example, people who use both alcohol and tobacco have a much higher risk of some cancers than would be expected if these two products were acting additively.
Tolerance	A range of values from 0 to 10 that indicates a species' ability to withstand organic waste. Species with 0 tolerance values are very intolerant of organic waste, while species with a 10 tolerance rating are very tolerant of organic waste. These values are based on numerous well-tested indices, most notably the Hilsenhoff species-level Biotic Index and the Family Biotic Index (Hilsenhoff, 1982).

REFERENCES

Allard, M., and G. Moreau. Influence of acidification and aluminum on the density and biomass of lotic benthic invertebrates. *Water Air Soil Pollut.* **1986**, *30*, 673–679.

Allard, M., and G. Moreau. Effects of experimental acidification on a lotic macroinvertebrate Community. *Hydrobiologia* **1987**, *144*, 37–49.

Amblard, C., P. Couture, and G. Bourdier. Effects of a pulp and paper mill effluent on the structure and metabolism of periphytic algae in experimental streams. *Aquat. Toxicol.* **1990**, *18* (3), 137–162.

Austin, A.P., G.E. Harris, and W.P. Lucey. Impact of an organophosphate herbicide glyphosate on periphyton communities developed in experimental streams. *Bull. Environ. Contamin. Toxicol.* **1991**, *47* (1), 29–35.

Bailey, H.C., C. Digiorgio, and D.E. Hinton. Impacts of multiple stressors on annual recruitment in striped bass (*Morone saxatilis*) in the Sacramento-San Joaquin estuary. *Marine Environ. Res.* **1996**, *42* (1–4),135.

Baker, J.P. and C.L. Schofield. Aluminum toxicity to fish in acidic waters. *Water Air Soil Pollut.* **1982**, *18*, 289–309.

Barbour, M.T. 1997. The re-invention of biological assessment in the U.S. *Hum. Ecol. Risk Assess.* **1997**, *3*, 933–940.

Barnthouse, L.W., D.R. Marmorek, and C.N. Peters. Assessment of multiple stressors at a regional scale. In *Multiple Stressors in Ecological Risk and Impact Assessment: Approaches to Risk Estimatio,* Ferenc, S. A. and J.A. Foran, Eds.; Society of Environmental Toxicology and Chemistry (SETAC): Pensacola, FL, 2000; pp. 1–26.

Belanger, S.E. Use of mesocosms in predicting ecosystem risk from cationic surfactant exposure. In *Predicting Ecosystem Risk*, J. Cairns, Jr., B.R. Neiderlehner, and D.B. Orvos, Eds.; Princeton Univ. Press: Princeton, NJ, 1992; Chap. 13, pp. 263–287.

Belanger, S. E. Literature review and analysis of biological complexity in model stream ecosystems: Influence of size and experimental design. *Ecotoxicol. Environ. Saf.* **1997**, *36*, 1–16.

Belanger, S.E., J.W. Bowling, D.M. Lee, E.M. LeBlanc, K.M. Kerr, D.C. McAvoy, S.C. Christman, and D.H. Davidson. Integration of aquatic fate and ecological responses to linear alkyl benzene sulfonate (LAS) in model stream ecosystems. *Ecotoxicol. Environ. Saf.* **2002**, *52*(2), 150–171.

Belanger, S.E., J.B. Guckert, J.W. Bowling, W.M. Begley, D.H. Davidson, E.M. LeBlanc, and D.M. Lee. Responses of aquatic communities to 25-6 alcohol ethoxylate in model stream ecosystems. *Aquat. Toxicol.* **2000**, *48,* 135–150.

Belanger, S.E., E.M. Meiers, and R.G. Bausch. Direct and indirect ecotoxicological effects of alkyl sulfate and alkyl ethoxysulfate on macroinvertebrates in stream mesocosms. *Aquat. Toxicol.* **1995**, *33*, 65–87.

Berrill, M., L. Hollett, A. Margosian, and J. Hudson. Variation in tolerance to low environmental pH by the crayfish *Orconectes rusticus*, *O. propinquus,* and *Cambarus robustus*. *Can. J. Zool.* **1985**, *63*, 2586–2589.

Bervoets, L., R. Blust, and R. Verheyen. Uptake of zinc by the midge larvae *Chironomus riparius* at different salinities: role of speciation, acclimation and aalcium. *Arch. Environ. Contam. Toxicol.* **1996**, *15*, 1423–1428.

Beyers D.W. Causal inference in environmental impact studies. *J. N. Am. Benthol. Soc.* **1998**, *17*, 367–373.

Blake, G. 1994. Are aquatic macrophytes useful in aquatic field tests. In *Freshwater Field Tests for Hazard Awareness of Chemicals*, I.R. Hill, F. Heimbach, P. Leeuwangh, and P. Matthiessen, Eds.; Lewis Publishers: Chelsea, Mich., 1994; pp. 183–190.

Boët, P., J. Belliard, R. Berrebi-dit-Thomas, and E. Tales. Multiple human impacts by the city of Paris on fish communities in the Seine River basin, France. *Hydrobiologia* **1999,** *410*, 59–68.

Booth, C.E., D.G. McDonald, B.P. Simons, and C.M. Wood. Effects of aluminum and low pH on net ion fluxes andion balance in the brook trout (*Salvelinus fontinalis*). *Can. J. Fish. Aquat. Sci.* **1988,** *45*(9), 1563–1574.

Bosserman, R.W. Distribution of heavy metals in aquatic macrophytes from Okefenokee Swamp. *Symp. Biol. Hung.* **1985***, 29*, 31–40.

Bowling, J.W., G.J. Leversee, P.F. Landrum, and J.P. Giesy. Acute mortality of anthracene-contaminated fish exposed to sun light. *Aquat. Toxicol.* **1983**, 3(1), 79–90.

Boyce, R.P. and S.B. Yamada. Effects of a parasite, *Eubothrium salvelini* (Cestoda: Pseudophyllidea), on the resistance of juvenile sockeyesSalmon, *Oncorhynchus nerka*, to zinc. *J. Fish. Res. Board Can.* **1977**, *50*, 706–709.

Brassard, C.A., A.L. Buikema, Jr., T.A. Bailey, M.J. Frankenberry, R.M. Lee, A.M. Stavola, C.L. Stunkard, and L.W. Touart. 1994. EPA aquatic mesocosm study review. In R.L.Graney, J.H. Kennedy, and J.H. Rodgers, Jr., eds., *Aquatic Mesocosm Studies in Ecological Risk Assessment*. SETAC Special Publication. Lewis, Boca Raton, FL, USA, pp. 17–24.

Breitburg, D.L., J.G. Sanders, C.C. Gilmour, C.A. Hatfield, R.W. Osman, G.F. Riedel, S.P. Seitzinger, and K.G. Sellner. Variability in responses to nutrients and trace elements, and transmission of stressor effects through an estuarine food web. *Limnol. Oceanogr.* **1999,** *44* (3 [2]), 837–863.

Britson, C.A., and S.T. Threlkeld. 1998. Abundance, metamorphosis, developmental, and behavioral abnormalities in *Hyla chrysoscelis* tadpoles following exposure to three agrichemicals and methyl mercury in outdoor mesocosms. *Bull. Environ. Contamin. Toxicol.* **1998**, *61(*2), 154–161.

Broderius, SJ. Modeling the joint toxicity of xenobiotics to aquatic organisms: basic concepts and approaches. In *Aquatic Toxicology and Risk Assessment*: 14th volume, Mayes, MA; Barron, MG, Eds.; ASTM STP 1124; Philadelphia, PA: American Society for Testing and Materials: Philadelphia, PA, 1991; pp. 107–127

Brown, D.C. Interactive *Effects of Temperature and Nutrients on Periphyton Assemblages from a Low-pH South Carolina Stream,* DP-MS-81-47.; Univ. of South Carolina: Columbia, 1981.

Brown, D.J.A. Effect of calcium and aluminum concentrations on the survival of brown trout *(Salmo trutta*) at low pH. *Bull. Environ. Contamin. Toxicol.* **1983**, 30, 582–587.

Bruland, K.W., J.R. Donat, and D.A. Hutchinson. Interactive influences of bioactive trace metals on biological production in oceanic waters. *Limnol Oceanogr.* **1991,** 36, 1555–1577.

Burton, T.M., and J.W. Allan. Influence of pH, aluminum and organic matter on stream invertebrates. *Can. J. Fish. Aquat. Sci.* **1986,** *43*(6), 1285–1289.

Calamari, D., and J.S. Alabaster. An approach to theoretical models in evaluating the effects of mixtures of toxicants in the aquatic environment. *Chemosphere* **1980,** *9*(9), 533–538.

Campbell, P.G.C., and P.M. Stokes. Acidification and toxicity of metals to aquatic biota. *Can. J. Fish. Aquat. Sci.* **1985,** 42, 2034–2049.

Canadian Council of Ministers of the Environment (CCME). *A Framework for Ecological Risk Assessment: General Guidance.* Document PN1195. The National Contaminated Sites Remediation Program: Winnipeg, Manitoba, 1996.

Canfield, T.J., N.E. Kemble, W.G. Brumbough, F.J. Dwyer, C.G. Ingersoll, and J.F. Fairchild. Use of benthic invertebrate community structure and the sediment quality triad to evaluate metal contaminated sediment in the Upper Clark Fork River, Montana. *Environ. Toxicol. Chem.* **1994,** *13*(12), 1999–2012.

Carder, J.P., and K.D. Hoagland. Combined effects of alachlor and atrazine on benthic algal communities in artificial streams. *Environ. Toxicol. Chem.* **1998,** *17*(7), 1415–1420.

Carpenter, D.E., and R.S. Lunetta. Challenges in forecasting the long-term impacts of multiple stressors on a mid-Atlantic region, USA. *Environ. Toxicol. Chem.* **2000,** *19*(4), 1076–1081.

Carrick, H.J., and R.L. Lowe. Response of Lake Michigan benthic algae to in situ enrichment with Si, N and P. *Can. J. Fish. Aquat. Sci.* **1988,** *45*, 271–279.

Cash, K.J. 1995. Northern River Basins Study project report no. 45, assessing and monitoring aquatic ecosystem health: approaches using individual population and community/ecosystem measurements. Northern River Basins Study, Edmonton, Alberta.

Cash, K.J., W.N. Gibbons, K.R. Munkittrick, S.B. Brown, and J. Carey. Fish health in the Peace, Athabasca and Slave River systems. *J. Aquat. Ecosys. Stress Recov.* **2000**, *8*(1), 77–86.

Chapman, P.M., A. Arthur, M. Paine, J. Vanderleelie, A. Maynard, J. Downie, and S. Cross. Sediment Related Investigations Off the Macaulay and Clover Point Sewage Outfalls. Capital Regional District, Project No. 3/073-11 : University of Victoria, BC, Canada, 1992.

Chapman, P.M., F.Y. Wang, C. Janssen, G. Persoone, and H.E. Allen. Ecotoxicology of metals in aquatic sediments: binding and release, bioavailability, risk assessments and remediation. *Can. J. Fish. Aquat. Sci.* **1998,** *55*(10), 2221–2243.

Christensen, E.R., and P.A. Zielski. Toxicity of arsenic and PCB to a green alga (*Chlamydomonas*), *Bull. Environ. Contam. Toxicol.* **1980***, 25*(1),43–48.

Christman, V.D., J.R. Voshell, Jr., D.G. Jenkins, M.S. Rosenzweig, R.J. Layton, and A.L. Buikema Jr. 1994. Ecological development and biometry of untreated pond mesocosms. In *Aquatic Mesocosm Studies in Ecological Risk Assessment,* Graney R.L., J.H. Kennedy, and J.H. Rodgers Jr., Eds.; Boca Raton, FL: CRC Press, 1994; pp. 105–127.

Clark, J.R., and J.L. Noles. Contaminant effects in marine/estuarine systems: field studies and scaled simulations. In *Aquatic Mesocosm Studies in Ecological Risk Assessment,* Graney R.L., J.H. Kennedy, and J.H. Rodgers Jr., Eds.; Boca Raton, FL: CRC Press, 1994; pp. 47–60. (Also available from NTIS, Springfield, VA: PB94-155488).

Clements, F.E. *Plant Succession: Analysis of the Development of Vegetation.* Carnegie Institute of Washington Publications **1916**, *242*, 1–512.

Clements, F.E. Nature and structure of the climax. *J Ecol.* **1936**, *24*, 252–284.

Clements, W.H., D.S. Cherry, and J. Cairns, Jr. Impact of heavy metals on insect communities in streams: a comparison of observational experimental results. *Can. J. Fish. Aquat. Sci.* **1988,** *45* (11): 2017–2025.

Clements, W.H., D.S. Cherry, and J. Cairns, Jr. The influence of copper exposure on predator-prey interactions in aquatic insect communities. *Freshwater Biol.* **1989,** *21*(3). 483–488.

Connell, D.W.*Bioaccumulation of Xenobiotic Compounds*. CRC Press Inc.: Boca Raton, FlL, 1990.

Connell, D.B., J.B. Sanders, and G.F. Riedel. Effects of multiple stressors on estuarine ecosystems: the relationship between nutrient levels and trace metal uptake in bivalves and fish In *American Society of Limnology and Oceanography Conference: Limnology and Oceanography: Navigating into the Next Century* - Abstracts. February 1-5, 1999, Aquatic Sciences Meeting, Santa Fe, NM.

Corkum, L.D. Patterns of benthic invertebrate assemblages in rivers of northwestern North America. *Freshwater Biol.* **1989,** *21*, 191–205.

Cottingham, K.L., and S.R. Carpenter. Predictive indices of ecosystem resilience in models of north temperate lakes. *Ecology* **1994,** *75*(7), 2127–2138.

Croce, B., R.M. Stagg, and M. Smith. Effects of exposure to a combination of beta-naphthoflavone and resin acids on biotransformation enzymes in Atlantic salmon parr (*Salmo salar*). *Aquat. Toxicol.* **1995,** 32(2-3): 161–176.

Crossland, N.O., G.C. Mitchell, D. Bennett, and J. Mazted. Use of outdoor artificial streams to determine threshold toxicity concentrations for a petrochemical effluent. *Environ. Chem. Toxicol.***1992,** 11, 49–59.

Cuker, B.E. Field Experiment on the influences of suspended clay and phosphorus on the plankton of a small lake. *Limnol. Oceanogr.* **1987,** *32*(4): 840–847.

Culp, J.M., and C.L. Podemski. *Impacts of Contaminants and Nutrients in Bleached Kraft Mill Effluent on Benthic Insect and Periphyton Communities: Assessments Using Artificial Streams, Athabasca River, 1993 and 1994.* Northern River Basins Study Technical Report No. 92. SSC-R71-49/3-92E. Edmonton, AB, 1996; pp. 1–70.

Culp, J.M., K.J. Cash, and F.J. Wrona. Integrated assessment of ecosystem integrity of large northern rivers: the Northern Rivers Basins Study example. *J. Aquat. Ecosys. Stress Recov.* **2000a,** *8*, 1–5.

Culp, J.M., K.J. Cash, and F.J. Wrona. Cumulative effects assessment for the Northern River Basins Study. *J. Aquat. Ecosys. Stress Recov.* **2000b,** *8*, 87–94.

Culp, J.M., R.B. Lowell, and K.J. Cash. Integrating mesocosm experiments with field and laboratory studies to generate weight-of-evidence risk assessments for large rivers. *Environ. Toxicol. Chem.* **2000a,** *19* (4[2]): 1167–1173.

Culp, J.M., R.B. Lowell, and K.J. Cash. Integrating in situ community experiments with field studies to generate weight-of-evidence risk assessments for large rivers. *Environ. Toxicol. Chem.* **2000b,** *19*, 1167–1173.

Culp, J.M., C.L. Podemski, and K.J. Cash. Interactive effects of nutrients and contaminants from pulp mill effluents on riverine benthos. *J. Aquat. Ecosys. Stress Recov.* **2000,** 8, 67–75.

DeAngelis, D.L. Energy flow, nutrient cycling, and ecosystem resilience. *Ecology* **1980,** *61*(4), 764–771.

Dean-Ross, D. 1989. Interactive effects of zinc and pH on periphyton in artificial streams. In *Abstracts of the 89th Annual Meeting of the American Society for Microbiology,* May 14, 1989; New Orleans; ASM: Washington, DC, 1989; pp. 89–360.

Dean-Ross, D. Effects of combined treatments of zinc and pH on attached bacterial abundance and activity. *Can. J. Microbiol.* **1991,** *37* (7), 491–496.

DeNoyelles, Jr. F., S.L. Dewey, D.G. Huggins, and W.D. Kettle. *Aquatic Mesocosms in Ecological Effects Testing: Detecting Direct and Indirect Effects of Pesticides. Aquatic Mesocosms in Ecological Risk Assessment;* CRC Press Ltd.: Boca Raton, FL, 1994; pp. 577–603.

Doe, K.G., W.R. Ernst, W.R. Parke, G.R. Julien, and P.A. Hennigar. 1988. Influence of pH on the acute lethality of fenitrithion, 2,4-D, and aminocarb and some pH altered sublethal effects of aminocarb on rainbow trout. *Can. J. Fish. Aquat. Sci.* 45(2): 287–293.

Draggan, S. Interactive Effect of Chromium Compounds and a Fungal Parasite on Carp Eggs. *Bull. Environ. Contam. Toxicol.* **1977**, *17*(6), 653–659.

Dube, M.G., J.M. Culp, and J.A. Smith. 1998. Development and Testing of a Mesocosm Technology for Environmental Effects Monitoring (EEM) Methods Application (pp. 43–46). Canadian Technical Report of Fisheries and Aquatic Sciences, Workshop to Assess Alternatives to the Fish Survey Component of the Environmental Effects Monitoring Program for Canadian Pulp and Paper Mills; Department of Fisheries and Oceans, Moncton, New Brunswick, Canada. **1998**, 43-46.

Dyer, S. D. and S. E. Belanger. Determination of the sensitivity of macroinvertebrates in stream mesocosms through field-derived assessments. *Environ. Chem. Toxicol.***1999,** *18*(12),2903–2907.

Dyer, S.D., C. White-Hull, G.J. Carr, E.P. Smith, and X. Wang. Bottom-up and top-down approaches to assess multiple stressors over large geographic areas. *Environ. Toxicol. Chem* **2000,** *19*(4), 1066–1075.

Ekelund R. Bioaccumulation and biomagnification of hydrophobic persistent compounds as exemplified by hexachlorobenzene. In *Chemicals in the Aquatic Environment ñ Advanced Hazard Assessment*, L. Landner, Ed.; Springer-Verlag: Berlin, Germany, 1989; pp. 128–149.

Environment Canada. *Pulp and Paper Technical Guidance Document for Aquatic Environmental Effects Monitoring.* EEM/1998/1, Environment Canada, Ottawa, Ontario, Canada, April 1998.

Environment Canada (D.S. Jeffries, Ed.). *1997 Canadian Acid Rain Assessment Volume 3: The Effects on Canada's Lakes, Rivers and Wetlands*. Environment Canada, Burlington, ON, 1997.

European Economic Community. Technical guidance documents in support of: The commission directive 93/67/EEC on risk assessment for new notified substances and the commission regulation (EC) 1488/94 on risk assessment for existing substances. Brussels, Belgium, 1994.

European Inland Fisheries Advisory Commission (EIFAC). *Water Quality Criteria for European Freshwater Fish. Revised Report on Combined Effects on Freshwater Fish and Other Aquatic Life of Mixtures of Toxicants in Water. Revision 1,* EIFAC Technical Paper 7732-18-5; European Inland Fisheries Advisory Commission: Rome, 1987.

Fairbrother, A., and R.S. Bennett. Multivariate statistical applications for addressing multiple stresses in environmental risk assessments. In *Multiple Stressors in Ecological Risk and Impact Assessment: Approaches to Risk Estimation,* Ferenc S.A. and J.A. Foran, Eds.; SETAC Press: Pensacola, FL, 2000; pp. 67–116.

Fairchild, J.F., T. Boyle, W.R. English, and C. Rabeni. Effects of sediment and contaminated sediment on structural and functional components of experimental stream ecosystems. *Water Air , Soil Pollut,* **1987**, *36*(3-4), 271–293.

Farag, A.M., C.J. Boese, D.F. Woodward, and H.L. Bergman. Physiological changes and tissue metal accumulation in rainbow trout exposed to food borne and water borne metals. *Environ. Toxicol. Chem.* **1994,** *13,* 2021–2029.

Federal Insecticide, Fungicide, and Rodenticide Act (7 U.S.C. s/s 136 et seq. 1996).

Finlayson, B.J. and K.M. Verrue. Toxicity of copper, zinc and cadmium to juvenile Chinook salmon. *Trans. Am. Fish. Soc.* **1982,** *111,* 645–650.

Fischer, R. Simulated or actual field testing: a comparison. In: *Aquatic Mesocosm Studies in Ecological Risk Assessment,* Graney R.L., J.H. Kennedy, and J.H. Rodgers Jr., Eds.; Boca Raton, FL: CRC Press, 1994; pp. 35–46.

Folke, C., C.S. Holling, and C. Perrings. Biological diversity, ecosystems and the human scale. *Ecol. Appl.* **1996**, *6*(4): 1018–1024.

Folsom, B.R., N.A. Popescu, and J.M. Wood. Comparative study of aluminum and copper transport and toxicity in an acid-tolerant freshwater green alga. *Environ. Sci. Technol.* **1986**, , 616–620.

Folt, C.L., C.Y. Chen, M.V. Moore, and J. Burnaford. Synergism and antagonism among multiple stressors. *Limnol. Oceanogr.* **1999**, *44*(3[2]), 864–877.

Forbes, V.E. Response of *Hydrobia ventrosa* (Montagu) to environmental stress: effects of salinity fluctuations and cadmium exposure on growth. *Funct. Ecol.* **1991**, *5*(5), 642–648.

Forsythe, B.L., and S.J. Klaine. The interactions of sulfate and selenate effects on brine shrimp *Artemis* sp. *Chemosphere* **1994**, *29*(4): 789–790.

Fox, G.A. Practical causal inference for ecoepidemiologists. *J. Toxicol. Environ. Health*, **1991**, *33*, 359–379.

Freeman, R.A., and W.H. Everheart. Toxicity of aluminum hydroxide complexes in neutral and basic media to rainbow trout. *Trans. Am. Fish. Soc.* **1971**, *4*, 644–658.

Furnival, G. M.; and Wilson, R. W. Regression by leaps and bounds. *Technometrics* **1974,** *16,* 499-511.

Furse, M.T., D. Moss, J.F. Wright, and P.D. Armitage. The influence of seasonal and taxonomic factors on the ordination and classification of running-water sites in Great Britain and on the prediction of their macro-invertebrate communities. *Freshwater Biol.* **1984**, *14*, 257–280.

Gauch, H.G., Jr. *Multivariate Analysis in Community Ecology*. Cambridge University Press: Cambridge, U.K., 1982.

Genter, R.B. Benthic algal populations respond to aluminum, acid, and aluminum-acid mixtures in artificial streams. *Hydrobiologia* **1995**, *306*(7), 19.

Genter, R.B., and D.J. Amyot. Freshwater benthic algal population and community changes due to acidity and aluminum-acid mixtures in artificial streams. *Environ. Toxicol. Chem.* **1994**, *13*(3), 369–380.

Gerhardt, A. 1990. Effects of sub-acute doses of cadmium on pH-stressed *Leptophlebia-marginata* l. and *Baetis-rhodani* Pictet Insecta Ephemeroptera. *Environ. Pollut.* **1990**, *67*(1), 29–42.

Gerhardt, A. Review of impact of heavy metals on stream invertebrates with special emphasis on acid conditions. *Water Air Soil Pollut.* **1993**, *66*(3-4), 289–314.

Gerritsen, J., and B. Jessup. *Effects of Multiple Stressors on Aquatic Ecosystems* (Report 96-IRM-2); Water Environment Research Foundation: Alexandria, VA, **2004**.

Gibbons, W.D., and K.R. Munkittrick. A sentinel monitoring framework for identifying fish population responses to industrial discharges. *J. Freshwater Ecosys. Health.* **1994**, *3*, 227–237.

Gobas, F.A.P.C., J.B. Wilcockson, R.W. Russell, and G.D. Haffner. Mechanisms of biomagnification in fish under laboratory and field conditions. *Environ. Sci. Technol.* **1999,** *34* (6), 980-987.

Golder Associates Ltd. *Phase II Moira River study: impacts of the former Deloro Mine site on the Moira River system.* Ontario Ministry of the Environment, Ontario, **2001**. Report available online at: http://www.ene.gov.on.ca/envision/deloro/moirastudy.htm.

Graney, R.L., J.H. Kennedy, and J.H. Rodgers Jr., Eds. *Aquatic Mesocosm Studies in Ecological Risk Assessment;* CRC Press: Boca Raton, FL, **1984**.

Grippo, R.S. and W.A. Dunson. The body ion loss biomarker: interactions between trace metals and low pH in reconstituted coal mine polluted water. *Environ. Chem. Toxicol.* **1996,** *15*(11), 1955–1963.

Growns, I.O., D.A. Pollard, and P.C. Gehrke. Changes in river fish assemblages associated with vegetated and degraded banks, upstream of and within nutrient-enriched zones. *Fish. Manag. Ecol.* **1998,** *5*(1), 55–69.

Guanguo, L. Different types of ecosystem experiments. In *Enclosed Experimental Marine Ecosystems: A Review and Recommendations,* C.M. Lalli, Ed.; Springer-Verlag: New York, 1990; pp. 7–19.

Guckert, J.B., D.D. Walker, and S.E. Belanger. Analytical chemistry of stream mesocosms exposed to dodecyl alkyl sulfate reveals rapid biodegradation. *Environ. Chem. Toxicol.* **1996,** *15*(3), 262–269.

Gunderson, L. Resilience, flexibility and adaptive management – antidotes for spurious certitude? *Conserv. Ecol.* **1996,** *3*(1). 7.

Haley, R.K., T.J. Hall, and T.M. Bousquet. Effects of biologically treated bleached-Kraft mill effluent before and after mill conversion to increased chlorine dioxide substitution: results of an experimental streams study. *Environ. Toxicol. Chem.* **1995,** 14(2), 287–298.

Hall, R.J., C.T. Driscoll, and G.E. Likens. Importance of hydrogen ions and aluminum in regulating the structure and function of stream ecosystems: an experimental test. *Freshwater Biol.* **1987,** *18,* 17–43.

Hall, T.J., R.K. Haley, and L.E. LaFleur. Effects of biologically treated bleached-Kraft mill effluent on cold water stream productivity in experimental stream channels. *Environ. Toxicol. Chem.* **1991,** *10*(8), 1051–1060.

Hall, R.I., P.R. Leavitt, R. Quinlan, A.S. Dixit, and J.P. Smol. Effects of agriculture, urbanization and climate on water quality in the northern Great Plains. *Limnol. Oceanogr.* **1999,** *44*(3[2]), 739–756.

Hanazato, T. Effects of repeated application of carbaryl on zooplankton communities in experimental ponds with or without the predator chaoborus. *Environ. Pollut.* **1991,** *74*(4), 309–324.

Hansen, L.D., K.J. Mair, and A.W. Knight. Effect of sulfate on the bioconcentration of selenate by *Chironomus decorus* and *Daphnia magna*. *Arch. Environ. Contam. Toxicol.* **1993,** *25*, 72–78.

Hansen, J.A., J.C.A. Marr, J. Lipton, D. Cacela, and H.L. Bergman. Differences in neurobehavioral responses of Chinook salmon (*Oncorhynchus mykiss*) exposed to copper and cobalt behavioral avoidance. *Environ. Toxicol. Chem.* **1998,** *18*(9), 1972–1978.

Hartwell, S.I., D.S. Cherry, and J. Cairns, Jr. Field validation of avoidance of elevated metals by fathead minnows *Pimephales-promelas* following in-situ acclimation. *Environ. Toxicol. Chem.* **1987,** *6*(3), 189–200.

Harwell, M.A. and J.H. Gentile. Environmental decision-making for multiple stressors: framework, tools, case studies and prospects. In *Multiple Stressors in Ecological Risk and Impact Assessment: Approaches to Risk Estimation*, SETAC Press: Pensacola, FL, 2000; pp. 169–226.

Hatakeyama, S. Effect of copper and zinc on the growth and emergence of *Epeorus-latifolium* (Ephemeroptera) in an indoor model stream. *Hydrobiologia* **1989,** *174*(1), 17–28.

Hay M.E. Defensive synergisms? Reply to Pennings. Ecology 77:1950-1953. *In* Folt, C.L., C.Y. Chen, M.V. Moore, and J. Burnaford. Synergism and antagonism among multiple stressors. *Limnol. Oceanogr.* **1999**, *44*(3[2]), 864–877.

Hay, M.E., Q.E. Kappel, and W. Fenical. Synergisms in plant defense against herbivores: interactions of chemistry, calcification and plant quality. *Ecology* **1994,** *75,* 1714–1727.

Hibachi, F., W. Pflueger, and H.T. Ratte. Use of small artificial ponds for assessment of hazards to aquatic ecosystems. *Environ. Toxicol. Chem.* **1992,** *11*(1), 27–34.

Hibachi, F., J. Berndt, and W. Pfluege. Fate and biological effects of an herbicide on two artificial pond ecosystems of different size. In *Aquatic Mesocosm Studies in Ecological Risk*

Assessment, Graney R.L., J.H. Kennedy, and J.H. Rodgers Jr., Eds.; Boca Raton, FL: CRC Press, SETAC Special Publications Series, 1993;(pp. 303ñ32).

Herbrandson, C., S.P. Bradbury, and D.L. Swackhamer. New testing apparatus for assessing interactive effects of suspended solids and chemical stressors on plankton invertebrates. *Environ. Chem. Toxicol.* **1999,** *18*(4), 679–684.

Hermanutz, R.O., J.G. Eaton, and L.H. Mueller. Toxicity of endrin and malathion mixtures to flagfish (*Jordanella floridae*). *Arch. Environ. Contam. Toxicol.* **1985,** *14*, 307–314.

Hilsenhoff, W.L. Using a biotic index to evaluate water quality in streams. Technical Bulletin no. 132. Department of Natural Resources: Madison, WI, 1982.

Hoagland, K.D., R.W. Drenner, J.D. Smith, and D.R. Cross. Freshwater community responses to mixtures of agricultural pesticides: effects of atrazine and bifenthrin. *Environ. Toxicol. Chem.* **1993,** *12*(4), 627–637.

Horne, M.T., and W.A. Dunson. The interactive effects of low pH, toxic metals, and DOC on a simulated temporary pond community. *Environ. Pollut.* **1995,** *89* (2), 155–161.

Hoss, D.E., L.C. Coston, and D.R. Colby. Synergistic effects of exposure to temperature and chlorine on survival of young-of-the-year estuarine fishes. In *Physiological Responses of Marine Biota to Pollutants*. Academic Press: New York, 1977, pp. 345–355.

Hossain, M.D., D.B. Ghatak, and S.K. Konar. Acute toxicity of mixture of a non-ionic detergent ekaline FI and an organic pesticide DDVP to fish, plankton and worm. *Environ.Ecol.* **1987,** *5*, 778–781.

Huey, D.W., M.C. Wooten, L.A. Freeman, and T.L. Beitinger. Effects of pH and chloride on nitrite-induced lethality in bluegill (*Lepomis macrochirus*). *Bull. Environ. Contam. Toxicol.* **1982,** *28*, 3–6.

Hughes, G.M., S.F. Perry, and V.M. Brown. A morphometric study of effects of nickel, chromium and cadmium on the secondary lamellae of rainbow trout gills. *Water Res.* **1979,** *13*, 665–679.

Hurlbert, S.H. Pseudoreplication and the design of field experiments. *Ecol. Monogr.* **1984,** *54*, 187–211.

Hutchinson, N.J., and J.B. Sprague. Toxicity of trace metal mixtures to American flagfish (*Jordanella floridae*) in soft, acidic water and implifications for cultural acidification. *Can. J. Fish. Aquat. Sci.* **1986,** *43*, 647–655.

Ince, N.H., N. Dirilgen, I.G. Apikyan, G. Tezcanli, and B. Ustun. Assessment of toxic interactions of heavy metals in binary mixtures: a statistical approach. *Arch. Environ. Contam. Toxicol.* **1999,** *36*, 365–72.

Jak, R.G., J.L. Maas, and M.C.T. Scholten. Ecotoxicity of 3,4-dichloroaniline in enclosed freshwater plankton communities at different nutrient levels. *Ecotoxicology* **1998,** 7(1),49–60.

Jarvinen, A.W., and G.T. Ankley. *Linkage of Effects to Tissue Residues: Development of a Comprehensive Database for Aquatic Organisms Exposed to Inorganic and Organic Chemicals.* SETAC Press: Pensacola, FL, **1999**.

Johnson, Philip C., J.H. Kennedy, R.G. Morris, F.E. Hambleton, and R.L. Graney. **1994**. Fate and effects of cyfluthrin (pyrethroid insecticide) in pond mesocosms and concrete microcosms. In: *Aquatic Mesocosm Studies in Ecological Risk Assessment,* Graney R.L., J.H. Kennedy and J.H. Rodgers Jr., Eds.; Society of Environmental Toxicity and Chemistry, Michigan: Lewis Publishers, pp. 337-371.

Karr, J.R., L.A. Toth, and D.R. Dudley. Fish communities of midwestern rivers: a history of degradation. *BioScience* **1985,** *35*(2), 90–95.

Kass, G.V. "An Exploratory Technique for Investigating Large Quantities of Categorical Data", Applied Statistics, Vol. 29, pp.119-127, **1980**.

Kemble, N.E., W.G. Brumbaugh, E.L. Brunson, F.J. Dwyer, C.G. Ingersoll, D.P. Monda, and D.F. Woodwards. Toxicity of metal contaminated sediments from the Upper Clark Fork River, Montana, to aquatic invertebrates and fish in laboratory exposures. *Environ. Chem. Toxicol.* **1994,** *13*(12), 1985–1997.

Kersting, K. and P.A. Van Wijngaarden. Effects of a pulsed treatment with the herbicide afalon (active ingredient Linuron) on macrophyte-dominated mesocosms. I. Responses of ecosystem metabolism. *Environ. Toxicol. Chem.* **1999,** *18*, 2859–2865.

Khatami S.H., D. Pascoe, and M.A. Learner. The acute toxicity of phenol and un-ionized ammonia, separately and together, to the Ephemeropteran *Baetis rhodani* (Pictet). *Environ. Pollut.* **1998,** *99*, 379–388.

Kiffney, P.M., and W.H. Clements. Effects of heavy metals on a macro-invertebrate assemblage from a Rocky Mountain stream in experimental microcosms. *J. N. Am. Benthol. Soc.,* **1994**, *13*, 511–523.

Kinross, J.H., P.A. Read, and N. Christofi. The influence of pH and aluminum on the growth of filamentous algae in artificial streams. *Arch. Hydrobiol.* **2000,** 149, 67–86.

Klein-MacPhee, G., B.K. Sullivan, and A.A. Keller. *Using Mesocosms to Assess the Influence of Food Resources and Toxic Materials on Larval Fish Growth and Survival.* In Water Quality & the Early Life Stages of Fishes, American Fisheries Society Symposium No. 14. Fuiman, L.A., American Society, Bethesda, Md. **1993**, 105-116.

Kosinski, R.J. Artificial streams in ecotoxicological studies. In A*quatic Toxicology: Fundamental Concepts and Methodologies,* A. Boudou and F. Ribeyre, Eds.; Vol. 1; CRC Press: Boca Raton, FL, 1989; pp. 298–316.

Kownacki, A., E. Dumnicka, E. Grabacka, B. Kawecka, and A. Starzecka. Stream ecosystems in mountain grassland (West Carpathians) Poland 14. The use of the experimental stream method in evaluating the effect of agricultural pollution. *Acta Hydrobiol.* **1985,** *27*(3), 381–400.

Krieger, K.A., D.B. Baker, and J.W. Kramer. Effects of herbicides on stream Aufwuchs productivity and nutrient uptake. *Arch. Environ. Contam. Toxicol.* **1988,** *17*(3), 299–306.

Kwak, H.I., M.O. Bae, M.H. Lee, Y.S. Lee, B.J. Lee, K.S. Kang, C.H. Chae, H.J. Sung, J.S. Shin, J.H. Kim, W.C. Mar, Y.Y. Sheen, and M.H. Cho. Effects of nonylphenol, bisphenol-a, and their mixture on the viviparous swordtail fish (*Xiphophorus helleri*). *Environ. Chem. Toxicol.* **2001,** *20*(4), 787–795.

Lambert, C.D. and C.A. Oviatt. **1986**. Manual of biological and geochemical techniques in coastal areas. MERL Series Report No. 1, Second Edition. University of Rhode Island, Kingston, RI. *In* Klein-MacPhee, G., B.K. Sullivan, and A.A. Keller. *Using Mesocosms to Assess the Influence of Food Resources and Toxic Materials on Larval Fish Growth and Survival.* In Water Quality & the Early Life Stages of Fishes, American Fisheries Society Symposium No. 14; American Fisheries Society: Bethesda, MD 1993, pp. 105–116.

Lamberti, G.A. and A.D. Steinman, Eds. Research in artificial streams: applications, uses, and abuses. *J. N. Am. Benthol. Soc.* **1993,** 2, 313–384.

Lamers, L.P.M., S. Van Roozendaal, and J.G.M. Roelofs. Acidification of freshwater wetlands: combined effects of non-airborne sulfur pollution and desiccation. *Water Air Soil Pollut.* **1998,** *105*(1-2), 95–106.

Landis, W.G. Uncertainty in the extrapolation from individual effects to impacts upon landscapes. *Hum. Ecol. Risk Assess.* **2002,** *8*(1): 193–204.

Landis, W.G., and J.A. Wiegers. Design considerations and a suggested approach for regional and comparative ecological risk assessment. *Hum. Ecol. Risk Assess.* **1997,** *3*(3), 287–297.

Landis, W.G., L.R. Bodensteiner, and J.F. McLaughlin. The relative Risk Rank Model for regional-scale risk assessment and potential applications in freshwater and terrestrial ecosystems.

In *Risk Assessment of Environmental Endpoints. Proceedings of a Workshop,* Lewis, G.D., N.G. Thom, J.E. Hay and K. Sukhia, Eds.; University of Aukland: Auckland, 1998; pp. 54–67.

Landis, W.G., M. Luxon, and L.R. Bodensteiner. Design of a Relative Risk Model Regional-Scale Risk Assessment with Confirmational Sampling for the Williamette and McKenzie Rivers, Oregon. In *Ninth Symposium on Environmental Toxicology and Risk Assessment: Recent Achievements in Environmental Fate and Transport. ASTM STP 1381,* Price, F.T., K.V. Brix, and N.K. Lane, Eds.; American Society for Testing and Materials: West Conshohocken, PA, 2000.

Landis, W.G., A.J. Markiewicz, R.A. Matthews, and G.B. Matthews. A test of the community conditioning hypothesis: persistence of effects in model ecological structures dosed with jet fuel JP-8. *Environ. Toxicol. Chem.,* **2000,** *19*(2), 327–336.

Landis, W.G. and J.F. McLaughlin. Design criteria and derivation of indicators for ecological position, direction and risk. *Environ. Toxicol. Chem.* **2000,** *19*(4): 1059–1065.

La Point, T.W., and J.F. Fairchild. Use of mesocosm data to predict effects in aquatic ecosystems: limits to interpretation. In *Aquatic Mesocosm Studies in Ecological Risk Assessment,* Graney R.L., J.H. Kennedy, and J.H. Rodgers Jr., Eds.; Boca Raton, FL: CRC Press, 1994; pp. 241–255 (SETAC Special Publications Series).

La Point, T.W., S.M. Melanson, and M.K. Morris. Relationships among observed metal concentrations, criteria and benthic community structural responses in 15 streams. *J. Water Pollut. Control Fed.,* **1984,** 56, 1030–1038.

Laursen, A., S. Seitzinger, R. Dekorsey, J. Sanders, and D. Breitburg. Multiple stressors in an estuarine system: effects of nutrients, trace metals, and biotic complexity on benthic photosynthesis and respiration. *Estuaries*, **1999**, *25* (1), 57–69.

Leaner, J.J., and R.P. Mason. The effect of thiolate organic compounds on methylmercury accumulation and redistribution in sheepshead minnows. *Environ. Chem. Toxicol.* **2001,** *20*(7), 1557–1563.

Lee, D.M., J.B. Guckert, S.E. Belanger, and T.C.J. Feijtel. Seasonal temperature declines do not decrease periphytic surfactant biodegradation or increase algal species sensitivity. *Chemosphere* **1997,** 35, 1143–1160.

Lehtinen, K.J., B. Axelsson, K. Kringstad and L. Stromberg. Characterization of pulp mill effluents by the model ecosystem technique. *Nord. Pulp Paper Res. J.* **1991**, 6:81-88. *In* Belanger, S. E. Literature review and analysis of biological complexity in model stream ecosystems: Influence of size and experimental design. *Ecotoxicol. Environ. Saf.* **1997**, *36*, 1–16.

Lehtinen, K.J., A. Larsson, and G. Klingstedt. Physiological disturbances in rainbow trout, *Salmo gairdneri* (R.), exposed at two temperatures to effluents from a titanium dioxide industry. *Aquat. Toxicol.* **1984,** *5*, 155–166.

Leland, H.V., S.V. Fend, T.L. Dudley, and J.L. Carter. Effects of copper on species composition of benthic insects in a Sierra Nevada, California, stream. *Freshwater Biol.***1989,** 21, 163–180.

Lemly, A.D. Modification of benthic insect communities in polluted streams combined effects of sedimentation and nutrient enrichment. *Hydrobiologia,* **1982,** *87*(3), 229–246.

Lewis, W.M., Jr. Investigations of the Relationship between Biodiversity and Ecosystem Function by Examination. NSF Dissertation Research, University of Colorado: Boulder, CIRES (Coop. Inst. for Research in Environmental Sciences), 1999.

Lin, C.K. and M.S. Simmons. Effects of pentachlorobiphenyl on growth of nutrient enriched phytoplankton from Lake Michigan. *J. Great Lakes Res.* **1981,** 7(4), 481–485.

Linden, O., J.R. Sharp, R. Laughlin, Jr., and J.M. Neff. Interactive effects of salinity, temperature and chronic exposure to oil on the survival and developmental rate of embryos of the estuarine killifish *Fundulus heteroclitus. Marine Biol.* **1979,** *10*(5), 101–109.

Logan, D.T. and H.T. Wilson. An ecological risk assessment method for species exposed to contaminant mixtures. *Environ. Toxicol. Chem.* **1995,** *14*(2): 351–359.

Lowe, R.L., J.B. Guckert, S.E. Belanger, D.H. Davidson, and D.W. Johnson. An evaluation of periphyton community structure and function on tile and cobble substrata in experimental stream mesocosms. *Hydrobiologia* **1996,** *328*, 135–146.

Lowell, R.B., and J.M. Culp. Cumulative effects of multiple effluent and low dissolved oxygen stressors on mayflies at cold temperatures. *Can. J.Fish. Aquat. Sci.* **1999,** *56*(9):1624–1630.

Lowell, R.B., J.M. Culp, and F.J. Wrona. Stimulation of increased short-term growth and development of mayflies by pulp mill effluent. *Environ. Toxicol. Chem.* **1995a,** *14*(9), 1529–1541.

Lowell, R.B., J.M. Culp, and F.J. Wrona. Toxicity testing with artificial streams: effects of differences in current velocity. *Environ. Toxicol. Chem.* **1995b,** *14*(7), 1209–1217.

Lowell, R.B., J.M. Culp, and M.G. Dube. A weight-of-evidence approach for northern river risk assessment: integrating the effects of multiple stressors. *Environ. Toxicol. Chem.* **2000,** *19*(4, Part 2), 1182–1190.

Lugthart, G.J. and J.B. Wallace. Effects of disturbance on benthic functional structure and production in mountain streams. *J. N. Am. Benthol. Soc.* **1992,** *11,* 138–164.

MacFarlane, R.D., G.L. Bullock, and J.J.A. McLauglin. Effects of five metals on susceptibility of striped bass to *Flexibacter columnaris*. *Am. Fish. Soc.* **1986,** *115*, 227–231.

Marchant, R., L.A. Barmuta, and B.C. Chessman. Preliminary study of the ordination and classification of macroinvertebrate communities from running waters in Victoria, Australia. *Austral. J. Mar. Freshwater Res.* **1994,** *45*, 945–962.

Marchant, R., L. Metzling, A. Graesser, and P. Suter. The organization of macroinvertebrate communities in the major tributaries of the LaTrobe River, Victoria, Australia. *Freshwater Biol.* **1985,** *15,* 315–331.

Marr, J.C.A., J.A. Hansen, J.S. Meyer, D. Cacela, T. Podrabsky, J. Lipton, and H.L. Bergman. Toxicity of cobalt and copper to rainbow trout: application of a mechanistic model for Predicting Survival. *Aquat. Toxicol.* **1998,** 43, 225–238.

Marshall, J.S., D.L. Mellinger, and J.I. Parker. Combined effects of cadmium and zinc on a Lake Michigan USA Canada zoo-plankton community. *J. Great Lakes Res.* **1981,** *7*(3), 215–223.

Massachusetts Weight-of-Evidence Workgroup. 1995. A weight-of-evidence approach for evaluating ecological risks. Draft Report. p. 6. Available online at: www.state.ma.us/dep/ors/files/weightev.pdf

Matthews, R.A., W.G. Landis, and G.B. Matthews. The community conditioning hypothesis and its application to environmental toxicology. *Environ. Toxicol. Chem.* **1996,** *15*(4), 597–603.

Mayasich, J.M., E.P. Karlander, and D.E. Terlizzi, Jr. Growth responses of *Nannochloris oculata* Droop and *Phaeodactylum tricornutum* Bohlin to the herbicide atrazine as influenced by light intensity and temperature in unialgal and bialgal assemblage. *Aquat. Toxicol.* **1987,** 10, 187–197.

McDonald, D.G. The interaction of environmental calcium and low pH on the physiology of the rainbow trout, Sa*lmo gairdneri* I. branchial and renal net ion H+ fluxes. *J. Exp. Biol.* **1983a,** 102, 123–140.

McDonald, D.G. The effects of H ions upon the gills of freshwater fish. *Can. J. Zool.* **1983b,** 61, 691–703.

McDonald, D.G., and C.L. Milligan. Sodium transport in the brook trout, *Salvelinus fontinalis*: effects of prolonged low pH exposure in the presence and absence of aluminum. *Can. J. Fish. Aquat. Sci.* **1988,** *45*,1606-1613.

McDowell, W.H., F.N. Scatena, and D.A. Schaefer. Effects of multiple disturbances on tropical stream chemistry. In *Conference: 48th Annual Meeting of the North American Benthological Society.* Keystone, Colorado, **2000,** *28* (5).

McFarlane, G.A., and W.G. Franzin. An Examination of cadmium, copper and mercury concentration in livers of Northern Pike, *Esox lucius* and white sucker, *Catostomus commersoni* from 5 lakes near a base metal smelter at Flin Flon, Manitoba, Canada. *Can. J. Fish. Aquat. Sci.* **1980,** *37*(10), 1573-1578.

Miersch, J., F. Barlocher, I. Bruns, and G.J. Krauss. Effects of cadmium, copper, and zinc on growth and thiol content of aquatic hyphomycetes. *Hydrobiologia* **1997,** 346, 77–84.

Mills, E.L., J.H. Leach, J.T. Carlton, and C.L. Secor. Exotic species and the integrity of the Great Lakes: lessons from the past. *BioScience* **1994,** *44* (10), 666–676.

Mitchell, G.C. System design for field tests in flowing waters: design and use of outdoor artificial streams in ecotoxicology. In *Freshwater Field Tests for Hazard Assessment of Chemicals,* Hill, I.R., F. Hibachi, P. Leeuwangh, and P. Mattiessen, Eds.; Lewis Publishers, Inc., Boca Raton, FL, 1994; pp. 127–139.

Moore, D.R.J., and S.M. Bartell. Estimating ecological risks of multiple stressors: advanced methods and difficult issues. In *Multiple Stressors in Ecological Risk and Impact Assessment: Approaches to Risk Estimation,* Ferenc, S.A., and J.A. Foran, Eds.; SETAC Press: Pensacola, FL, 2000; pp. 117–168).

Moss, D., M.T. Furse, J.F. Wright, and P.D. Armitage. The prediction of the macro-invertebrate fauna of unpolluted running-water sites in Great Britain using environmental data. *Freshwater Biol.* **1987,** *17*, 41–52.

Mount, D.R., C.G. Ingersoll, D.D. Gulley, J.D. Fernandez, T.W. La Point, and H.L. Bergnman. Effect of long-term exposure to acid, aluminum, and low calcium on adult brook trout (*Salvelinus fontinalis*). 1. Survival, growth, fecundity and progeny survival. *Can. J. Fish. Aquat. Sci.* **1988,** *45*(9), 1623–1632.

Mount, D.R., J.R. Hockett, and W.A. Gern. Effect of long-term exposure to acid, aluminum, and low calcium on adult brook trout (*Salvelinus fontinalis*). 2. Vitellogenesis and osmoregulation. *Can. J. Fish. Aquat. Sci.* **1988,** 45(9),1633–1642.

Mount, D.R., A.K. Barth, T.D. Garrison, K.A. Barten, and J.R. Hockett. Dietary and waterborne exposure of rainbow trout (*Oncorhynchus mykiss*) to copper, cadmium, lead and zinc using a live diet. *Environ. Toxicol. Chem.* **1994,** *13*(12): 2031–2041.

Munkittrick, K.R., and D.G. Dixon. A holistic approach to ecosystem health assessment using fish population characteristics. *Hydrobiologia* **1989,** 188/189, 123–135.

Munkittrick, K.R. and M.E. McMaster. Effects-driven assessment of multiple stressors using fish populations. In *Multiple Stressors in Ecological Risk and Impact Assessment: Approaches to*

Risk Estimation, Ferenc, S.A. and J.A. Foran, Eds.; SETAC Press: Pensacola, FL, 2000; pp. 27–66.

Munkittrick, K.R., M.A. McMaster, K. Van Der Kraak, C. Portt, W.N. Gibbons, A. Farwell, and M. Gray. *Developments of Methods for Effects-Driven Cumulative Effects Assessment Using Fish Populations: Moose River Project*; SETAC Press: Pensacola, FL, 2000.

Murray, C.L. Multiple Toxicity of Two Pesticides and Two Metals to *Mysidopsis bahia* (chlorpyrifos, chlordane, copper sulfate, cadmium chloride). M.S. Thesis, University of Houston-Clear Lake, 1997.

Naddy, R., T.W. La Point, and S.J. Klaine. Toxicity of arsenic, molybdenum and selenium combinations to *Ceriodaphnia dubia*. *Environ. Toxicol. Chem.* **1995,** 14(2), 329–336.

National Research Council Canada. *Acidification in the Canadian aquatic environment: scientific criteria for assessing the effects of acidic deposition on aquatic ecosystems.* NRCC No. 18475. Ottawa, Ontario, 1981.

Neubert, M.G., and H. Casell. Alternatives to resilience for measuring the responses of ecological systems to perturbations. *Ecology* **1996,** *78*(3), 653–665.

Neville, C.M. Physiological response of juvenile rainbow trout *Salmo gairdneri*, to acid and aluminum—prediction of field responses from laboratory data. *Can. J. Fish. Aquat. Sci.* **1985,** *42,* 2004–2019.

Newman, R.M., and J.A. Perry. The combined effects of chlorine and ammonia on litter breakdown in outdoor experimental streams. *Hydrobiologia* **1989,** *184*(1-2), 69–78.

Nicholls, K.H., and G.J. Hopkins. Recent changes in Lake Erie (North Shore) phytoplankton: cumulative impacts of phosphorus loading reductions and the zebra mussel introduction. *J. Great Lakes Res.* **1993,** *19*(4), 637–647.

Nishimoto, M., BTL Eberhart, H.R. Sanborn, C. Krone, U. Varanasi, and J.E. Stein. Affects of a complex mixture of chemical contaminants on hepatic glutathione, l-cysteine and (-glutamylcysteine synthetase in English sole. *Environ. Toxicol. Chem.* **1995,** *14*(3), 461–469.
Niyogi, D.K., D.M. McKnight, and W.M. Lewis. Influences of water and substrate quality for periphyton in a montane stream affected by acid mine drainage. *Limnol. Oceanogr.* **1999,** *44*(3, Part 2), 804–809.

Northern River Basins Study Board. Northern River Basins Study, report to the Ministers, 1006. Northern River Basins Study, Edmonton, Alberta, 1996.

Obery, A.M., and W.G. Landis. A regional multiple stressor risk assessment of the Colorus Creek watershed applying the relative risk model. *Hum. Ecol.l Risk Assess.* **2002,** 8(2), 405–428.

Odum, E.P., J.T. Finn, and E.H. Franz. Perturbation theory and the subsidy-stress gradient. *BioScience* **1979,** *29* (4), 349–352.

Odum, E.P. Trends expected in stressed ecosystems. *Bioscience,* **1985,** *37,* 419–422.

Ormerod, S.J., and R.W. Edwards. The ordination and classification of macroinvertebrate assemblages in the catchment of the River Wye in relation to environmental factors. *Freshwater Biol.* **1987,** 17, 533–546.

Osman. R.W., J.G. Sanders, D.L. Breitburg, M.H. Bundy, G.F. Riedel, C.C.C. Gilmour, and S.P. Seitzinger. Effects of multiple stresses on estuarine ecosystem dynamics in large-scale field enclosures. In *American Society of Limnology and Oceanography Conference: Limnology and Oceanography: Navigating into the Next Century,* abstracts; February 1-5, 1999 Aquatic Sciences Meeting, Santa Fe, NM; 1999.

Paulson, C., S. Reeves, S.P. Canton, and T. Moore. A*bility to Discriminate Habitat versus Chemical Limitations: User Guidance Manual* (98-WSM-1); Water Environment Research Foundation: Alexandria, VA, 2001.

Peckarsky, B.L. Biotic interactions or abiotic limitations? a model of lotic community structure. In *Dynamics of Lotic Ecosystems,* Bartell, S.M., and T.D. Fontaine, III, Eds.; Ann Arbor Science Publishers, Inc.: Ann Arbor, MI, 1983; pp. 131–160.

Perry, J.A., and N.H. Troelstrup, Jr. Whole ecosystem manipulation a productive avenue for test system research. *Environ. Toxicol. Chem.* **1988**, *7*(11), 941–951.

Podemski, C.L., and J.M. Culp. Nutrient and contaminant effects of bleached Kraft mill effluent on benthic algae and insects of the Athabasca River. In *Environmental Fate and Effects of Pulp and Paper Mill Effluents,* M. K. Servos, J. H. Carey, and G. J. Van Der Kraak, Eds.; St. Lucie Press: Delray Beach, FL, 1996; pp. 571–580.

Pratt, J.R. and J. Cairns, Jr. 1996. Ecotoxicology and the redundancy problem: understanding effects on community structure and function. In *Ecotoxicology: a Hierarchical Treatment,* Newman M.C., and C.H. Jagoe, Eds.; Lewis Publishers: Boca Raton, FL, 1996; pp. 347–370.

Proulx, M., F.R. Pick, A. Mazumder, P.B. Hamilton, and DRS Lean. Experimental evidence for interactive impacts of human activities on lake Algal species richness. *Oikos,* **1996,** *76*(1), 191–195.

Quinn, J.M. and C.W. Hickey. Characterisation and classification of benthic invertebrate communities in 88 New Zealand rivers in relation to environmental factors. *New Zealand J. Mar. Freshwater Res.* **1990a,** *24,* 387–409.

Quinn, J.M., and C.W. Hickey. Magnitude of effects of substrate particle size, recent flooding, and catchment development on benthic invertebrates in 88 New Zealand rivers. *New Zealand J. Mar. Freshwater Res.* **1990b** *24,* 411–427.

Rachlin, J.W., and A. Grosso. The growth response of the green alga chlorella vulgaris to combined divalent cation exposure. *Arch. Environ. Contam. Toxicol.* **1993,** 24, 16–20.

Rapport, D.J., and W.G. Whitford. How ecosystems respond to stress. *BioScience* **1999,** *49,* 193–204.

Rasmussen, K., and C. Lindegaard. Effects of iron compounds on macroinvertebrate communities in a Danish lowland river system. *Water Res.* **1988,** 22, 1101–1108.

Raufflet, E. Linking social and ecological systems: management practices and social mechanisms for building resilience. *Conserv. Ecol.* **1998,** *4*(2), 5. Available online at: http://www.consecol.org/vol4/iss2/art5.

Regal, R.R. and S.J. Lozano. 1994. Optimal design of aquatic field studies. In *Aquatic Mesocosm Studies in Ecological Risk Assessment,* Graney R.L., J.H. Kennedy, and J.H. Rodgers Jr., Eds.; Boca Raton, FL: CRC Press, 1994; pp. 157–171 (SETAC Special Publications Series).

Reid. S.D. and D.G. McDonald. 1988. Effects of Cd, Cu, and Low pH on Ion Fluxes in the Rainbow Trout. *Can. J. Fish. Aquat. Sci.,* 45: 244–253.

Reynoldson, T.B., R.C. Bailey, K.E. Day, and R.H. Norris. Biological guidelines for fresh-water sediment based on benthic assessment of Sediment (the BEAST) using a multivariate approach for predicting biological state. *Austral. J. Ecol.* **1995,** 20(1), 198–219.

Rhodes L., E. Casillas, B. McKnight, W. Gronlund, M. Meyers, O.P. Olson, and B. McCain. Interactive effects of cadmium, polychlorinated biphenyls, and fuel oil on experimentally exposed English sole (*Parophrys vetulus*). *Can. J. Fish. Aquat. Sci.* **1985,** *42*, 1870–1880.

Richards J.G., B.K. Burnison, and R.C. Playle. Natural and commercial dissolved organic matter protects against the physiological effects of a combined cadmium and copper exposure on rainbow trout. *Can. J. Fish. Aquat. Sci.* **1999,** *56*(3), 407–418.

Rodgers, John H., N.O. Crosslands, E.R. Kline, W.B. Gillespie, Jr., R.A. Figuera, and P.B. Dorn. Design and construction of model stream ecosystems. *Ecotoxicol. Environ. Saf,* **1996,** *33*, 30–37.

Rodgers, Jr. D., S. Cherry, and R.K. Guthrie. Cycling of elements in duckweed (*Lemna perpusilla*) in an ash settling basin and swamp drainage. *Water Res.* **1978,** *12*, 765–770.

Rosenberg, D.M., and V.H. Resh, Eds. *Freshwater Biomonitoring and Benthic Invertebrates;* Chapman & Hall: New York, 1993.

Rutt, G.P., N.S. Weatherly, and S.J. Ormerod. Relationships between the physicochemistry and macroinvertebrates of British upland streams: the development of modeling and indicator systems for predicting fauna and detecting acidity. *Freshwater Biol.* **1990,** *24*, 463–480.

Sandberg, R.S., and W.G. Landis. Persistence of the effects of Jet-A in a microcosm with releases from the sediment. *Environ. Toxicol. Chem.* **2001,** *20*(9), 1942–1950.

Sanders, J.G., M.G. Weinstein, and D.L. Breitburg. *Effects of multiple stressors on estuarine ecosystems: Seasonal Variability in Response of Phytoplankton Communities to Trace Elements.* Conference: Limnology and Oceanography: Navigating into the Next Century. American Society of Limnology and Oceanography. February 1-5, 1999 Santa Fe, NM.**1999**, 27 (4).

Sarkar, U.K., and S.K. Konar. Interactions of insecticide thiodan (endosulfan) and heavy metal chromium and impact on fish and aquatic ecosystem. *Chem. Ecol.***1997,** *13*(4), 271–278.

Schalk, G. Single and interactive effects of low pH and leech predators on frog embryos and adults *(Macrobdella decora, Rana sylvatica, R. clamitans, R. catesbeiana, R. pipiens)*. M.Sc. Thesis, Carleton University, Canada, 1997.

Schamberger, M, A.H. Farmer and J.W. Terrell. *Habitat suitability index models: introduction.* U.S.D.I. Fish and Wildlife Service, 1982, FWS/OBS-82/10.

Schindler D.W. Detecting ecosystem responses to anthropogenic stress. *Can. J. Fish. Aquat. Sci.* **1987,** *44* (Suppl. 1), 6–25.

Schindler, D.W. Sustaining aquatic ecosystems in boreal regions. *Conserv. Ecol.* **1998,** *2*(2), 18. Available online at: http://www.consecol.org/vol2/iss2/art18.

Schlueter, M.A., S. Guttaman, I. James, T. Oris, and A.J. Bailer. Differential survival of fathead minnows, *Pimephales promelas*, as affected by copper exposure, prior population stress, and allozyme genotypes. *Environ. Toxicol. Chem.* **1997,** *16*(5), 939–947.

Scott, B.F., P.J. Wade, and W.D. Taylor. Impact of oil and oil dispersant mixtures on the fauna of fresh water ponds. *Sci. Total Environ.***1984,** *35*(2), 191–206.

Shapiro, J., and G.E. Glass. Synergistic effects of phosphate and manganese on growth of lake superior algae. *Verhandlungen Int. Vereinigung Limnol.* **1975,** *19*(1), 395–404.

Sigmon, C.F. Influence of 2,4-D and 2,4,5-T on life history characteristics of *Chironomus* (Diptera: Chironomidae). *Bull. Environ. Contam. Toxicol.* **1979,** *21*, 596–599.

Smith, T.R., and T.A. Haines. Mortality, growth, swimming activity and gill morphology of brook trout (*Salvelinus fontinalis*) and Atlantic Salmon (*Salmo salar*) exposed to low pH with and without aluminum. *Environ. Pollut.* **1995,** *90*(1), 33–40.

Spehar, R.L. Cadmium and zinc toxicity to flagfish, *Jordanella floridae*. *J. Fish. Res. Board Can.* **1976,** *33*(9), 1939–1945.

Stephenson, R.R. Hazard assessment from field tests in the UK: pond mesocosms and microcosms—the need for a flexible approach to experiment design. In *Freshwater Field Tests for Hazard Assessment of Chemicals,* Hill, I.R., F. Hibachi, P. Leeuwangh, and P. Mattiessen, Eds.; CRC Press: Boca Raton, FL, 1994; pp. 439–448.

Stewart, A.R. Accumulation of Cd by a freshwater mussel is reduced in the presence of Cu, Zn, Pb and Ni. *Can. J. Fish. Aquat. Sci.* **1999,** 56(3), 467–478.

Stuijfzand, S.C., M.J. Jonker, E. van Ammelrooy, and W. Admiraal. Species-specific responses to metals in organically enriched river water, with emphasis on effects of humic acids. *Environ. Pollut.* **1999,** *106*(1), 115–121.

Stuijfzand, S.C., M. Helms, M.H.S. Kraak, and W. Admiraal. Interacting effects of toxicants and organic matter on the midge *Chironomus riparius* in polluted river water. *Ecotoxicol. Environ. Saf.* **2000,** *46*, 351–356.

Sullivan, J.F., G.H. Atchison, D.J. Kolar, and A.W. McIntosh. Changes in the predator-prey behavior of fathead minnows (*Pimephales promelas*) and largemouth bass (*Micropterus salmoides*) caused by cadmium. *J. Fish. Res. Board Can.* **1978,** *35*. 446–451.

Suter, G.W. A critique of ecosystem health concepts and indexes. *Environ. Toxicol. Chem.* **1993a,** *12*, 1533–1539.

Suter, G.W. *Ecological Risk Assessment*. Lewis Publishers, Boca Raton, FL, 1993b, pp. 373–375.

Suter, G.W. *Risk Characterization for Ecological Risk Assessment of Contaminated Sites*, ES/ER/TM-200; U.S. Department of Energy: Washington, DC, 1996.

Taub F.B. A biological model of a freshwater community – a gnotobiotic ecosystem. *Limnol. Oceanogr.* **1969**, *14* (1),136–142.

Taylor, B.R. Rapid assessment procedures: Radical re-invention or just sloppy science? *Hum.Ecol. Risk Assess.* **1997,** *3*, 1005–1016.

Thatcher, T.O., M.J. Schneider, and E.G. Wolf. Bioassays on the combined effects of chlorine, heavy metals, and temperature on fishes and fish food organisms: Part I. Effects of chlorine and temperature on juvenile brook trout (*Salvelinus fontinalis*). *Bull. Environ. Contam. Toxicol.* **1976,** *15*(1), 40–48.

Thompson, D.G., S.B. Holmes, D.G. Pitt, K.R. Solomon, and K.L. Wainio-Keize. Applying concentration-response theory to aquatic enclosure studies. In *Aquatic Mesocosm Studies in Ecological Risk Assessment,* Graney R.L., J.H. Kennedy, and J.H. Rodgers Jr., Eds.; Boca Raton, FL: CRC Press, **1994**; pp. 129–156 (SETAC Special Publications Series).

Tietge, J.E., R.D. Johnson and H.L. Bergman. Morphometric changes in gill secondary lamellae of brook trout (Salvelinus Fontinalis) after long-term exposure to acid and aluminum. *Can. J. Fish. Aquat. Sci.* **1988,** *45*(9): 1643–1648.

Tjeerdema, R.S., R.J. Kauten, and D.G. Crosby. Interactive effects of pentachlorophenol and hypoxia in the abalone (*Haliotis-rufescens*) as measured by in-vivo phosphorus-31 NMR spectroscopy. *Aquat. Toxicol.* **1991,** *21*(3-4), 279–294.

Tjeerdema, R.S., R.J. Kauten, and D.G. Crosby. Interactive effects of pentachlorophenol and temperature in the abalone (*Haliotis rufescens*) as measured by in vivo phosphorus-31 NMR spectroscopy. *Aquat. Toxicol.* **1993,** *26*(1/2), 117–132.

Touart, L.W. 1994. Regulatory endpoints and the experimental design of aquatic mesocosm tests. In *Aquatic Mesocosm Studies in Ecological Risk Assessment,* Graney R.L., J.H. Kennedy, and J.H. Rodgers Jr., Eds.; Boca Raton, FL: CRC Press, 1994; pp. 25–33 (SETAC Special Publications Series).

U.S. Environmental Protection Agency (U.S. EPA). *Estimating Concern Levels for Concentrations of Chemical Substances in the Environment,* U,S, EPA, Office of Pollution Prevention and Toxics, Health and Environmental Review Division (7403), Environmental Protection Agency: Washington, DC, 1984.

U.S. EPA. 1992. *Framework for Ecological Risk Assessment*, EPA/630/R-92/001; Risk Assessment Forum, U.S. EPA: Washington, DC, 1992.

U.S. EPA. *Ecological Risk Assessment Guidance for Superfund, Process for Designing and Conducting Ecological Risk Assessment, Interim Final*, EPA 540-R-97-006, OSWER Directive # 9285.7-25; U.S. EPA: Washington, DC, 1997.

U.S. EPA. *Guidelines for Ecological Risk Assessment.* EPA/630/R-95/002F; Risk Assessment Forum, U.S. EPA: Washington, DC, 1998.

U.S. EPA. *Stressor Identification Guidance Document*, EPA/822/B-00/025; Office of Research and Development, U.S.EPA: Washington, DC, 2000.

Van Der Gaag, M.A. 1992. Combined effects of chemicals: An essential element in risk extrapolation for aquatic ecosystems. In *Water Science and Technology*, Matsui, S. Ed.; Oxford, U.K. **1992**; pp. 441–447.

Van Der Geest, H.G., G.D. Greve, M. Boivin, M.H.S. Kraak and C.A.M. van Gestel. **2000**. Mixture toxicity of copper and diazinon to larvae of the mayfly (*Ephoron virgo*) judging additivity at different effect levels. *Environ. Chem. Toxicol,* 19(12): 2900–2905.

Van Gardinger, P.R., G.M. Foody, and P.J. Curran. *Scaling-Up. From Cell to Landscape.* Cambridge Univ. Press: Cambridge, U.K., 1997.

Walker, R.L., C.M. Wood, and H.L. Bergman. Effects of low pH and aluminum on ventilation in the brook trout (*Salvelinus fontinalis*). *Can. J. Fish. Aquat. Sci.* **1988**, *45* (9), 1614–1622.

Wang, W.X., and R.C.H. Dei. Effects of major nutrient additions on metal uptake in phytoplankton. *Environ. Pollut.* **2001**, *111*, 233–240.

Weigers, J.K., H.M. Feder, L.S. Mortensen, D.G. Shaw, V.J. Wilson, and W.G. Landis. A regional multiple-stressor rank-based ecological risk assessment for the fjord of Port Valdez, Alaska. *Hum. Ecol. Risk Assess.* **1998**, *4*(5), 1125–1173.

Whitford, W.G. Desertification: implications and limitations of the ecosystem health metaphor. In *Evaluating and Monitoring the Health of Large-Scale Ecosystems,* Rapport, D.J., C. Gaudet and P. Calow, Eds.; Springer-Verlag: Heidelberg, Germany, 1995; pp. 273–293.

Wicklum, D., and R.W. Davies. Ecosystem health and integrity. *Can. J. Bot.* **1995**, *73*, 997–1000.

Wilbur, H.M. Regulation of structure in complex systems experimental temporary pond communities. *Ecology* **1987**, *68*(5), 1437–1452.

Williams, W.T. Air pollution disease in the California forests. *Environ. Sci. Technol.* **1980,** *14*, 179–182.

Wood, C.M., R.C. Playle, B.P. Simons, G.G. Goss, and D.G. McDonald. Blood gases, acid-base status, ions, and hematology in adult brook trout (*Salvelinus fontinalis*) under acid/aluminum exposure. *Can. J. Fish. Aquat. Sci.* **1988a**, *45*, 1575–1586.

Wood, C.M., D.G. McDonald, C.E. Booth, B.P. Simons, C.G. Ingersoll, and H.L. Bergman. Physiological evidence of acclimation to acid/aluminum stress in adult brook trout (*Salvelinus fontinalis*). 1. Blood composition and net sodium fluxes. *Can. J. Fish. Aquat. Sci.* **1988b**, *45*, 1587–1596.

Wood, C.M., B.P. Simons, D.R. Mount and H.L. Bergman. Physiological evidence of acclimation to acid/aluminum stress in adult brook trout (*Salvelinus fontinalis*). 2. Blood parameters by cannulation. *Can. J. Fish. Aquat. Sci.* **1988c**, *45*, 1597–1605.

Wood, C.M., D.G. Mc Donald, C.G. Ingersoll, D.R. Mount, O.E. Johannsson, S. Landsberger, and H.L. Bergman. Effects of water acidity, calcium and aluminum on whole body ions of brook trout continuously exposed from fertilization to swim up: a study by instrumental neutron activation analysis. *Can. J. Fish. Aquat. Sci.* **1990**, *47*, 1593–1603.

Woodward, D.F., A.F. Farag, H.L. Bergman, A.J. DeLonay, and E.E. Little. Metals contaminated benthic invertebrates in the Clark Fork River, Montana: effects on age-0 brown trout and rainbow trout. *Can. J. Fish. Aquat. Sci.* **1994**, *52*, 1994–2004.

Woodwell, G. M. The blue planet: of wholes and parts and man. In *Disturbance and Ecosystems*, H.A. Mooney and M. Godron, Eds.; Springer-Verlag: Berlin, 1983; pp. 2–10.

Wright, J.F., D. Moss, P.D. Armitage, and M.T. Furse. A preliminary classification of running-water sites in Great Britain based on macro-invertebrate species and the prediction of community type using environmental data. *Freshwater Biol.* **1984**, *14*, 221–256.

Zeeman, M.G. Ecotoxicity testing and estimating methods development under section 5 of the toxic substances control act (TSCA). In *Fundamentals of Aquatic Toxicology: Effects, Environmental Fate and Risk Assessment*, G.M Rand, Ed.; Taylor and Francis: Washington, DC, 1995; pp 703–716.

WASTEWATER UTILITY

Alabama
Montgomery Water Works & Sanitary Sewer Board

Alaska
Anchorage Water & Wastewater Utility

Arizona
Gila Resources

Glendale, City of, Utilities Department

Mesa, City of

Peoria, City of

Pima County Wastewater Management

Arkansas
Little Rock Wastewater Utility

California
Calaveras County Water District

Central Contra Costa Sanitary District

Contra Costa Water District

Crestline Sanitation District

Delta Diablo Sanitation District

Dublin San Ramon Services District

East Bay Dischargers Authority

East Bay Municipal Utility District

El Dorado Irrigation District

Fairfield-Suisun Sewer District

Fresno Department of Public Utilities

Irvine Ranch Water District

Las Virgenes Municipal Water District

Livermore, City of

Lodi, City of

Los Angeles, City of

Los Angeles County, Sanitation Districts of

Napa Sanitation District

Orange County Sanitation District

Palo Alto, City of

Riverside, City of

Sacramento Regional County Sanitation District

San Diego Metropolitan Wastewater Department, City of

San Francisco, City & County of

San Jose, City of

Santa Barbara, City of

Santa Rosa, City of

South Bayside System Authority

South Coast Water District

South Orange County Wastewater Authority

Sunnyvale, City of

Union Sanitary District

West Valley Sanitation District

Colorado
Boulder, City of

Colorado Springs Utilities

Littleton/Englewood Water Pollution Control Plant

Metro Wastewater Reclamation District, Denver

Connecticut
The Mattabasset District

New Haven, City of, WPCA

District of Columbia
District of Columbia Water & Sewer Authority

Florida
Broward, County of

Fort Lauderdale, City of

Gainesville Regional Utilities

JEA

Miami-Dade Water & Sewer Authority

Orange County Utilities Department

Orlando, City of

Reedy Creek Improvement District

Seminole County Environmental Services

St. Petersburg, City of

Stuart Public Utilities

Tallahassee, City of

Tampa, City of

Toho Water Authority

West Palm Beach, City of

Georgia
Atlanta Department of Watershed Management

Augusta, City of

Clayton County Water Authority

Cobb County Water System

Columbus Water Works

Fulton County

Gwinnett County Department of Public Utilities

Savannah, City of

Hawaii
Honolulu, City and County of

Illinois
American Bottoms Wastewater Treatment Plant

Greater Peoria Sanitary District

Kankakee River Metropolitan Agency

Metropolitan Water Reclamation District of Greater Chicago

North Shore Sanitary District

Wheaton Sanitary District

Iowa
Ames, City of

Cedar Rapids Wastewater Facility

Des Moines Metro Wastewater Reclamation Authority

Iowa City, City of

Kansas
Johnson County Unified Wastewater Districts

Unified Government of Wyandotte County/Kansas City, City of

Kentucky
Louisville & Jefferson County Metropolitan Sewer District

Louisiana
Sewerage & Water Board of New Orleans

Maine
Bangor, City of

Portland Water District

Maryland
Anne Arundel County Bureau of Utility Operations

Howard County Department of Public Works

Washington Suburban Sanitary Commission

Massachusetts
Boston Water & Sewer Commission

Upper Blackstone Water Pollution Abatement District

Michigan
Ann Arbor, City of

Detroit, City of

Holland Board of Public Works

Lansing, City of

Owosso Mid-Shiawassee County WWTP

Saginaw, City of

Wayne County Department of Environment

Wyoming, City of

Minnesota
Rochester, City of

Western Lake Superior Sanitary District

Missouri
Independence, City of

Kansas City Missouri Water Services Department

Little Blue Valley Sewer District

Nebraska
Lincoln Wastewater System

Nevada
Henderson, City of

New Jersey
Bergen County Utilities Authority

Passaic Valley Sewerage Commissioners

New York
New York City Department of Environmental Protection

Rockland County Solid Waste Management Authority/ Sewer District

North Carolina
Charlotte/Mecklenburg Utilities

Durham, City of

Metropolitan Sewerage District of Buncombe County

Orange Water & Sewer Authority

Ohio
Akron, City of

Butler County Department of Environmental Services

Columbus, City of

Metropolitan Sewer District of Greater Cincinnati

Northeast Ohio Regional Sewer District

Summit, County of

Oklahoma
Tulsa, City of

Oregon
Clean Water Services

Eugene/Springfield Water Pollution Control

Water Environment Services

Pennsylvania
Philadelphia, City of

University Area Joint Authority, State College

South Carolina
Charleston Commissioners of Public Works

Mount Pleasant Waterworks & Sewer Commission

Spartanburg Sanitary Sewer District

Tennessee
Cleveland, City of

Knoxville Utilities Board

Murfreesboro Water & Sewer Department

Nashville Metro Water Services

Texas
Austin, City of

Dallas Water Utilities

Denton, City of

El Paso Water Utilities

Fort Worth, City of

Gulf Coast Waste Disposal Authority

Houston, City of

San Antonio Water System

Trinity River Authority

Utah
Salt Lake City Corporation

Virginia
Alexandria Sanitation Authority
Arlington, County of
Fairfax County Virginia
Hampton Roads Sanitation District
Henrico, County of
Hopewell Regional Wastewater Treatment Facility
Loudoun County Sanitation Authority
Lynchburg Regional WWTP
Prince William County Service Authority
Richmond, City of
Rivanna Water & Sewer Authority

Washington
Edmonds, City of
Everett, City of
King County Department of Natural Resources
Seattle Public Utilities
Sunnyside, Port of
Yakima, City of

Wisconsin
Green Bay Metro Sewerage District
Madison Metropolitan Sewerage District
Milwaukee Metropolitan Sewerage District
Racine, City of
Sheboygan Regional Wastewater Treatment
Wausau Water Works

Australia
South Australian Water Corp.
Sydney Water Corp.
Water Corp. of Western Australia

Canada
Greater Vancouver Regional District
Toronto, City of, Ontario
Winnipeg, City of, Manitoba

Mexico
Servicios de Agua y Drenaje de Monterrey, I.P.D.

New Zealand
Watercare Services Limited

Singapore
Singapore Public Utilities Board

United Kingdom
Yorkshire Water Services Limited

STORMWATER UTILITY

California
Los Angeles, City of, Department of Public Works
Monterey, City of
San Francisco, City & County of
Santa Rosa, City of
Sunnyvale, City of

Colorado
Boulder, City of

Georgia
Griffin, City of

Iowa
Cedar Rapids Wastewater Facility
Des Moines Metro Wastewater Reclamation Authority

Kansas
Overland Park, City of

Kentucky
Louisville & Jefferson County Metropolitan Sewer District

Maine
Portland Water District

Minnesota
Western Lake Superior Sanitary District

North Carolina
Charlotte, City of, Stormwater Services

Pennsylvania
Philadelphia, City of

Tennessee
Chattanooga Storm Water Management

Washington
Seattle Public Utilities

STATE

Arkansas Department of Environmental Quality
Fresno Metropolitan Flood Control District, Calif.
Urban Drainage & Flood Control District, Colo.

CORPORATE

ADS Environmental Services
The ADVENT Group Inc.
Alan Plummer & Associates
Alden Research Laboratory
Aqua-Aerobic Systems, Inc.
Aquateam–Norwegian Water Technology Centre A/S
BaySaver Inc.
BioVir Laboratories, Inc.
Black & Veatch
Boyle Engineering Corporation
Brown & Caldwell
Burns & McDonnell
CABE Associates Inc.
The Cadmus Group
Camp Dresser & McKee Inc.
Carollo Engineers Inc.
Carpenter Environmental Associates Inc.
CDS Technologies Inc.
Chemtrac Systems Inc.
CH2M HILL
Damon S. Williams Associates, LLC
David L. Sheridan, P.C.
Dewling Associates, Inc.
Earth Tech Inc.
Eco-Matrix
Ecovation
Environmental Engineers International
EMA Inc.
The Eshelman Company Inc.
Finkbeiner, Pettis, & Strout Inc.
Freese & Nichols, Inc.
ftn Associates Inc.
Fuss & O'Neill, Inc.
Gannett Fleming Inc.
Golder Associates Inc.
Greeley and Hansen LLC
Hazen & Sawyer, P.C.
HDR Engineering Inc.
HNTB Corporation
HydroQual Inc.
Infilco Degremont Inc.
Ingersoll-Rand Energy Systems
Insituform Technologies Inc.
Institute for Environmental Technology & Industry, Korea
Jacobson Helgoth Consultants Inc.
Jason Consultants Inc.
Jordan, Jones, & Goulding Inc.
KCI Technologies Inc.
Kelly & Weaver, P.C.
Kennedy/Jenks Consultants
Komline Sanderson Engineering Corporation
Lawler, Matusky & Skelly Engineers, LLP
Limno-Tech Inc.
Lombardo Associates Inc.
Malcolm Pirnie Inc.
McKim & Creed
MEC Analytical Systems Inc.
Metcalf & Eddy Inc.
MWH
New England Organics
Odor & Corrosion Technology Consultants Inc. (OCTC)
Oswald Green, LLC
PA Government Services Inc.
Parametrix Inc.
Parsons
Post, Buckley, Schuh & Jernigan
R&D Engineering/Conestoga Rover & Associates
The RETEC Group
RMC, Inc. (Raines, Melton & Carella)
R.M. Towill Corporation
Ross & Associates Ltd.
Royce Technologies
SAIC Maritime Technical Group
Short Elliott Hendrickson, Inc.
Stantec Consulting Group, Inc.
Stormwater Management, Inc.
Synagro Technologies, Inc.
Tetra Tech Inc.
Trojan Technologies Inc.
URS Corporation
Veolia Water NATC
Wade-Trim Inc.
Weston Solutions Inc.
Woodard & Curran
WRc/D&B, LLC
WWETCO, LLC
Zoeller Pump Company

INDUSTRY

American Electric Power
ChevronTexaco Energy Research & Technology Company
The Coca-Cola Company
Dow Chemical Company
DuPont Company
Eastman Kodak Company
Eli Lilly & Company
Merck & Company Inc.
ONDEO Services
Procter & Gamble Company
PSEG Services Corp.
RWE Thames Water Plc
Severn Trent Services Inc.
United Water Services LLC

Note: List as of 7/1/04

WERF Product Order Form

As a benefit of joining the Water Environment Research Foundation, subscribers are entitled to receive one complimentary copy of all final reports and other products. Additional copies are available at cost (usually $10). To order your complimentary copy of a report, please write "free" in the unit price column. WERF keeps track of all orders. If the charge differs from what is shown here, we will call to confirm the total before processing.

Name ______ Title ______

Organization ______

Address ______

City ______ State ______ Zip Code ______ Country ______

Phone ______ Fax ______ Email ______

Stock #	Product	Quantity	Unit Price	Total
			Postage & Handling	
			VA Residents Add 4.5% Sales Tax	
			Canadian Residents Add 7% GST	
			TOTAL	

Method of Payment:

(All orders must be prepaid.)

❑ Check or Money Order Enclosed

❑ Visa ❑ Mastercard ❑ American Express

Account No. ______ Exp. Date ______

Signature ______

Shipping & Handling:

Amount of Order	United States	Canada & Mexico	All Others
Up to but not more than:	Add:	Add:	Add:
$20.00	$5.00*	$8.00	50% of amount
30.00	5.50	8.00	40% of amount
40.00	6.00	8.00	
50.00	6.50	14.00	
60.00	7.00	14.00	
80.00	8.00	14.00	
100.00	10.00	21.00	
150.00	12.50	28.00	
200.00	15.00	35.00	
More than $200.00	Add 20% of order	Add 20% of order	

*minimum amount for all orders

Note: Please make checks payable to the Water Environment Research Foundation.

To Order (Subscribers Only):

Log on to www.werf.org and click on the "Product Catalog."

Phone: (703) 684-2470
Fax: (703) 299-0742.

WERF
Attn: Subscriber Services
635 Slaters Lane
Alexandria, VA 22314-1177

To Order (Non-Subscribers):

Non-subscribers may be able to order WERF publications either through WEF (www.wef.org) or IWAP (www.iwapublishing.com). Visit WERF's website at www.werf.org for details.

IWA Publishing's authorised EU representative for General Product Safety Regulations is Diane D'Arras, 15 rue Duret, 75116 Paris, France, e-mail: safety@iwap.co.uk.

Printed and bound by CPI Group (UK) Ltd, Croydon, CR0 4YY
12/05/2026
02108876-0016